913
The Secret Wisdom
of Genesis

913

THE SECRET WISDOM OF GENESIS

by
Rabbi Yitzchak Ginsburgh

edited by
Rabbi Moshe Genuth

גלעיני
ואביטה
נפלאות
מתורתך

Gal Einai

The Torah and Mathematics Series

913: The Secret Wisdom of Genesis

Rabbi Yitzchak Ginsburgh

Edited by Rabbi Moshe Genuth

Printed in the United States of America and Israel
First Edition

Copyright © 5776 (2015) by Gal Einai

For information:

Israel: Gal Einai
         PO Box 1015
         Kfar Chabad 6084000
         tel. (in Israel): 1-700-700-966
         tel. (from abroad): 972-3-9608008

email: books@inner.org

Web: www.inner.org
Twitter: @RabbiGinsburgh

Printed with the support of the Torah Institute of Yeshivat Od Yosef Chai.

Layout: David Hillel
Cover design: Heni Ben Aharon

ISBN: 978-965-7146-96-5

"...נכון שיכתוב בצורת ספר השיעורים שלומד.
בברכה להצלחה..."

"...It would be proper to publish your classes
in book form.
With blessings for success..."

*– from a letter from the Lubavitcher Rebbe
to the author, Elul 5741*

TABLE OF CONTENTS

INTRODUCTION

Traditional mathematical analysis of Torah

Torah is no stranger to mathematics. There is a long tradition of Torah interpretation and commentary based on mathematical findings in the Torah text. The source of the mathematical analysis in the Torah can be identified in *gematria*, which constitutes one of the 32 principles with which the Torah is studied according to the compilation of these principles by the Second Temple era sage, Rabbi Eliezer (son of Rabbi Yossi Hagleelee). By the Middle Ages, we find two central commentaries on the Pentateuch, the *Roke'ach* and the *Ba'al Haturim*, whose core is based on *gematria*.

Textual and mathematical analyses

When considering these two texts from a modern perspective, it is important to differentiate between the techniques of textual analysis they employ and the level of mathematics they use to analyze them. Because it is considered part of the original tradition of Torah interpretation as received originally from Moshe Rabbeinu, the techniques of textual analysis provide us with a methodology that does not change and is not easily augmented. However, the mathematics used in these texts can certainly benefit from our more modern understanding. As it stands, apart from a few sporadic examples, the level of mathematics found in these early works is relatively rudimenatry, employing little more than basic addition operations.▸

▸ One particularly notable exception appears in another medieval legal text called the *Kol Bo*. In one of its final chapters, the anonymous author brings a beautiful phenomenon linking the Torah's Priestly Blessing with a variant of the Chinese Remainder Theorem.

The mathematization of Torah

Still, even though we can certainly employ more complex

ix

mathematics—reflecting our modern knowledge—to the textual analysis, the value of the findings documented by the early scholars lies not in the mathematical prowess they demonstrate, but rather in their spiritual content. Traditionally, the early scholars are considered to have written with the aid of *Ru'ach Hakodesh* (the holy spirit), considered a minor form of prophecy. While our own generation is considered lacking when it comes to the holiness required to attain a level of even minor prophecy, it is not uncommon to find Torah scholars who are well versed in even advanced mathematics.

It is in this respect that Rabbi Ginsburgh's star shines brightest. He is without any doubt the greatest innovator in Torah mathematics that the tradition has seen. However, Rabbi Ginsburgh has not only upgraded the mathematics employed in analyzing the Torah text. One can safely argue that he possesses a sensitivity to the Torah's quantitative dimension that is so strong that many times his mathematical findings seem to demonstrate a form of minor prophecy.

At the heart of the mathematics that Rabbi Ginsburgh uses to approach the Torah are the prime numbers and the figurate numbers. A word about each is due.

Prime numbers in mathematics

Prime numbers—numbers that are divisible by only 1 and themselves—are what make the fundamental theorem of arithmetic possible. This theorem states that every composite integer is a product of exactly one set of prime numbers. For instance, the number 15 is the product of 3 and 5; 20 is the product of 2, 2, and 5; 111 is the product of 3 and 37. Following the fundamental theorem of arithmetic, it can be stated that the prime numbers are to the integers what the atomic elements are to matter. To carry the analogy out, just as every molecule is composed of a certain number of each of the elemental atoms, so every integer is composed (and therefore called a composite number) out of a certain number of each of the prime numbers. For example, every molecule of water, or H_2O, is thus composed of 2 hydrogen atoms with 1 oxygen atom. Similarly, the number 90 is the product of 2, 3, 3, and 5, or one 2, two 3's, and one 5.

Prime numbers in Torah

Because prime numbers are the elements of the integers and all words in Hebrew have an integer value, it is revealing to look at the primes whose product is the value of a particular number or word. Going back to the example of 90, we can say that 90 has 30 measures of 3 in it, or 3 measures of 30; we could also say that 90 has 45 measures of 2, or vice versa; and so on. This leads us to identify the multiples of other central numbers as base measures as well. For example, the most important word in the Pentateuch is God's essential Name, *Havayah* (יהוה), whose value is 26. Though 26 is a composite number (2 times 13), it is still very interesting to look at its multiples—26, 52, 78, 104, 130, 156, 182, 208, etc. and view them as a group of inherently connected numbers (mathematically, we would call this the 0 modulus 26 group of integers, i.e., numbers whose remainder when divided by 26 is 0; in more advanced mathematics, these numbers are said to form a residue class).

Even the most rudimentary reading of the Arizal's Kabbalistic writings reveals that the analysis of Hebrew word's values based on base measures is central to his mystical intentions. One simple illustration is that one should dip bread (לֶחֶם) in salt (מֶלַח)—the value of both words is 78—three times before eating, because 78 is 3 times 26, meaning that as a base measure, God's essential Name, *Havayah*, fits into bread and salt 3 times.

It is worth noting that this second point—recognizing the multiples of certain important numbers that are not prime—is independent of the importance of the primes in arithmetic. Indeed, we do not find that the Arizal gave special importance to primes in his teachings, but as noted, other special numbers, notably 26, enjoy the limelight and serve as the basic measuring rods with which other numbers are analyzed. Yet, Rabbi Ginsburgh does afford the primes a special place alongside the more traditionally important multiples like 26. Of special significance are certain pairs of primes that appear together in the Torah, most notably 7 and 13 (another such pair is 23 and 37 whose relationship is discussed in various places in the present volume). Nevertheless, what Rabbi Ginsburgh normally does is to seek the prime factors of every composite number, given that the primes, as explained, constitute a sort of elemental key for analyzing every integer.

Summation of prime factors

Expanding upon the traditions found in earlier works on *gematria*, he has introduced a new form of analysis based on the prime numbers called the "foundation" (יְסוֹד) of a number. The foundation of an integer n is defined as the sum of its prime factors. Thus, the foundation of God's essential Name *Havayah*, whose value is 26, would be 15 (13 plus 2), alluding to the first two letters of the Name, *Kah* (יה), whose numerical value is 15, and indeed are considered the foundational letters of the full Name *Havayah*.

Enumerating the prime numbers

Because of their importance, it is revealing to enumerate the primes: the first prime number, the second prime number, and so on. Here we immediately encounter a relatively recent change of convention. For centuries, the number 1 was considered a prime number, the most prime of them all, if you will, for every prime can be divided only by itself and by 1. However, because of various considerations, in the past few decades, mathematicians have opted to segregate the number 1 in a class by itself, neither prime nor composite. Given that in the end the issue of whether 1 is a prime or not is a matter of convention (how one chooses one's definitions), it is equally valid to begin enumerating the primes with 1 or with 2 (provided that consistency is maintained).

For example, we noted above the importance of the pair of primes 7 and 13. One of the many mathematical properties relating this pair of numbers is that 13 is the 7th prime number (when enumerating 1 as the first prime).

Enumerating the prime numbers allows us to define another mathematical funcation of a word, which we call the "source" (מְקוֹר) of the word. The source is found by multiplying the ordinal values of the primes whose product is the word's value. Therefore, for example, to calculate the source of *Havayah*, God's essential Name, we factor its value 26 into primes: 2 and 13. Since 2 is the 2nd prime and 13, as noted, is the 7th, the source is their multiple, or $2 \cdot 7 = 14$.

Prime positions

Another revealing method of analysis is to look at the letters in prime positions in a verse or a group of verses. For instance, the Torah's first verse has 28 letters, so of these 10 are in prime positions (1, 2, 3, 5, 7, 11, 13, 17, 19, 23) and the rest, 18 letters, are in composite positions. Explicitly, the letters in the prime positions are in bold: בְּרֵאשִׁית בָּרָא אֱלֹהִים את הַשָּׁמַיִם וְאֵת הָאָרֶץ. The value of the letters in prime positions is 301, the value of the word "fire" (אֵשׁ), whose relationship with the Torah's first verse will be the extended topic of chapter 9 in this volume.

Figurate numbers

Figurate numbers are one of the most central innovations Rabbi Ginsburgh has introduced in learning Torah using mathematics. Figurate numbers are symmetrical figures that can be used to depict the integers. Perhaps the most familiar figurate numbers are the squares. We are all aware that the series of square numbers 1, 4, 9,

16, etc. can be drawn in the figure of a square number of points equal to the number. For example, the first few squares can be drawn as,

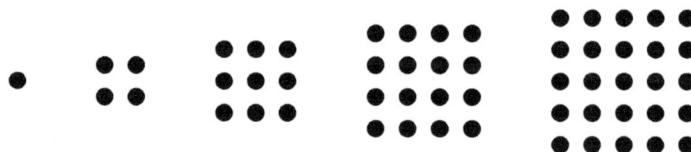

Of course the square numbers can be defined by the function $f[n] = n^2$.

Apart from the square numbers, many other figures depict other series of integers. Another basic form is that of the triangular numbers: 1, 3, 6, 10... Not surprisingly, the triangular numbers can be drawn as triangles of points. Thus, the first few triangular numbers are,

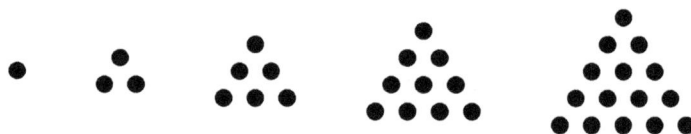

As can be seen from the figures above, the triangular numbers are simply the sum of integers from 1 to n. They are defined by the function:

$$\triangle n = \sum_{k=1}^{n} k$$

Another basic form is that of the interface (or inspirational) numbers: 1, 5, 13, 25,… These numbers actually have two different isomorphic forms in which they can be drawn, one being a 45-degree rotation of the other. The two figures of the interface numbers are thus,

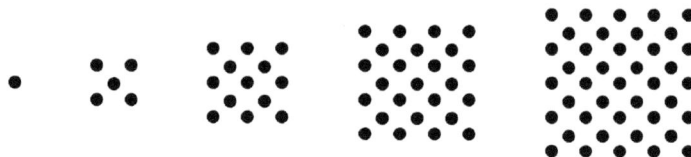

and,

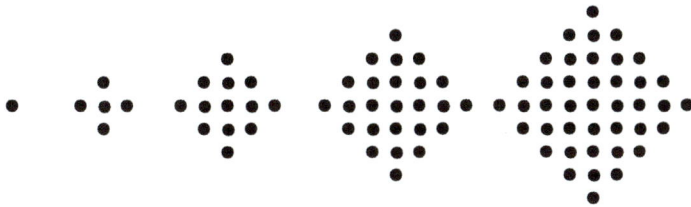

The interface numbers are actually the sum of two consecutive square numbers. Thus the equation defining them is $f[n] = n^2 \perp (n-1)^2$.

An exhaustive treatment of all the types of figurate numbers is beyond our scope here, but following is a chart of some of the major types used in Torah analysis:

figurate number	notation	function
triangle	$\triangle n$	$\triangle n = \dfrac{n(n+1)}{2}$
square	n^2	$n^2 = n \cdot n$
interface (inspirational)	$\boxdot n$	$\boxdot n = n^2 + (n-1)^2$
Eve	$\mathsf{x}n$	$f[n] = 2\triangle n - 1$
diamond	$\blacklozenge n$	$\blacklozenge n = 2\triangle n$
covenant	$\mathsf{I}n$	$\mathsf{I}n = 2\triangle n + 1$
Shabbat (hexagonal)	$\bigcirc n$	$\bigcirc n = 6\triangle n + 1$
Magen David	$\maltese n$	$\maltese n = 12\triangle n + 1$
Chashmal (pentagonal)	$\cap n$	$\cap n = \triangle(n-1) + n^2$
house	$\text{A}n$	$\text{A}n = \triangle n + n^2$
cubic	n^3	$n^3 = n \cdot n \cdot n$
tetrahedron	$\triangle n$	$\triangle n = \displaystyle\sum_{k=0}^{n} \triangle k$

Square numbers represent consummate unity

How are figurate numbers used in understanding Torah? Historically, one of the most important findings in this respect from the past 100 years was first described by Tuvia Wexler. He noted that when various passages of Torah text are combined in the liturgy with original liturgical texts, in many cases the total number of letters or words would be a square number. Case in point is the Friday night *Amidah* (composed in the 2nd Temple period), where the passage describing Shabbat in Genesis (2:1 to 2:3) and containing 144 letters (12^2) was combined with the original liturgical passage beginning "You sanctified" (אַתָּה קִדַּשְׁתָּ), containing 81 letters (9^2). Together these two passages contain 225 or 15^2 letters!

When this finding was presented to Rabbi Ginsburgh, he immediately recognized its significance and connection with a central Kabbalistic principle known as inter-inclusion (הִתְכַּלְלוּת). In Kabbalah, square numbers indicate a state known as inter-inclusion, whereby a perfect unification is attained between two or more things. For example, when the masculine and the feminine unify perfectly, the unification includes 4 elements (4 is the square of 2): the masculine in the masculine, the feminine in the masculine, the masculine in the feminine, and the feminine in the feminine. Another example of inter-inclusion can be found in the Counting of the Omer—the 49 days from the second day of Passover to *Shavu'ot*. 49 is the square of 7, and the days of the Omer are understood as a unification between the 7 emotive faculties, from loving-kindness to kingdom. This perfect unification, or inter-inclusion, requires each of the faculties to unify with each of the others (and with itself, as in the masculine in the masculine and the feminine in the feminine in the previous example). Therefore, 49 such unifications are required.

An example of square numbers

Once the significance of the square numbers is understood in the context of inter-inclusion, their symbolic value in Torah becomes clear and we can state that when we find that two or more verses are joined together by the Torah's structure, or by the interpretations offered by the sages, and their total sum of letters (or words) equals a square number, that is an indication of their essential unity. One especially revealing example of this principle can be seen in the three verses describing the Almighty's 13 Measures of Mercy (יג מִדּוֹת הָרַחֲמִים) revealed to Moses following the sin of the Golden Calf. The three verses in Hebrew are:

וַיֵּרֶד יְ־הוה בֶּעָנָן וַיִּתְיַצֵּב עִמּוֹ שָׁם וַיִּקְרָא בְשֵׁם יְ־הוה. וַיַּעֲבֹר יְ־הוה עַל פָּנָיו וַיִּקְרָא יְ־הוה יְ־הוה אֵל רַחוּם וְחַנּוּן אֶרֶךְ אַפַּיִם וְרַב חֶסֶד וֶאֱמֶת. נֹצֵר חֶסֶד לָאֲלָפִים נֹשֵׂא עָוֹן וָפֶשַׁע וְחַטָּאָה וְנַקֵּה לֹא יְנַקֶּה פֹּקֵד עֲוֹן אָבוֹת עַל בָּנִים וְעַל בְּנֵי בָנִים עַל שִׁלֵּשִׁים וְעַל רִבֵּעִים.

The sages recognize the beginning of the 13 Measures of Mercy in the two instances of God's essential Name, *Havayah.*▸ There is no clear indication that this should be the case when reading the verses. Likewise, they identify the final measure of Divine mercy in the word וְנַקֵּה, and again, there is nothing in the verse that would indicate that this word is distinct in any way.

However, all 3 verses together contain exactly 169 letters,▸▸ first of all indicating that they form a unified and coherent unity. Of course, 169 is the square of 13, providing a beautiful allusion to the sages' identification that these verses describe God's 13 Measures of Mercy. We are of course motivated then to draw the verses in a square form, like so:

▸ Let us put in bold the actual words included in the 13 Measures of Divine Mercy:

וַיֵּרֶד יְ־הוה בֶּעָנָן וַיִּתְיַצֵּב עִמּוֹ שָׁם וַיִּקְרָא בְשֵׁם יְ־הוה. וַיַּעֲבֹר יְ־הוה עַל פָּנָיו וַיִּקְרָא **יְ־הוה יְ־הוה אֵ־ל רַחוּם וְחַנּוּן אֶרֶךְ אַפַּיִם וְרַב חֶסֶד וֶאֱמֶת. נֹצֵר חֶסֶד לָאֲלָפִים נֹשֵׂא עָוֹן וָפֶשַׁע וְחַטָּאָה וְנַקֵּה** לֹא יְנַקֶּה פֹּקֵד עֲוֹן אָבוֹת עַל בָּנִים וְעַל בְּנֵי בָנִים עַל שִׁלֵּשִׁים וְעַל רִבֵּעִים.

The actual text of the Measures of Mercy consists of 67 letters. However, there are also exactly 67 letters surrounding the actual text—20 letters from the beginning of the second verse, to their beginning (וַיַּעֲבֹר יְ־הוה עַל פָּנָיו וַיִּקְרָא) and another 47 letters from their end to the termination of the third verse (לֹא יְנַקֶּה פֹּקֵד עֲוֹן אָבוֹת עַל בָּנִים וְעַל בְּנֵי בָנִים עַל שִׁלֵּשִׁים וְעַל רִבֵּעִים). This number, 67, is the value of "understanding" (בִּינָה).

▸▸ The first and second verse together have 91 letters, the triangle of 13. The third verse has 78 letters, the triangle of 12. One of the most fundamental properties of triangular numbers is that the sum of two consecutive triangles is equal to the square of the larger, or, $n^2 = \triangle n + \triangle(n-1)$.

In this case, the sum of 91 and 78 is 169, the square of 13.

ו	ן	נ	ע	ב	ה	ו	ה	ה	י	ד	ר	י	ו
ק	י	ו	ם	ש	ו	מ	ע	ב	צ	י	ת	י
ב	ע	י	ו	ה	ו	ה	י	ם	ש	ב	א	ר
י	ו	ו	י	נ	פ	ל	ע	ה	ו	ה	י	ר
ל	א	ה	ו	ה	י	ה	ו	ה	י	א	ר	ק
א	ך	ר	א	ן	ו	נ	ח	ו	ם	ו	ח	ר
ת	מ	א	ו	ד	ס	ח	ב	ר	ו	ם	י	פ
נ	ם	י	פ	ל	א	ל	ד	ס	ח	ר	צ	נ
א	ט	ח	ו	ע	ש	פ	ו	ן	ו	ע	א	ש
ק	פ	ה	ק	נ	י	א	ל	ה	ק	נ	ו	ה
י	נ	ב	ל	ע	ת	ו	ב	א	ן	ו	ע	ד
ל	ע	ם	י	נ	ב	י	נ	ב	ל	ע	ו	ם
ם	י	ע	ב	ר	ל	ע	ו	ם	י	ש	ל	ש

Once the verses are drawn in this new form, we reveal many wonderful things about them that could not have been seen otherwise. For instance, we find that the letters in the square's four corners are וו ש ם, whose sum is 352, the exact value of the Measure of Mercy "patient" (אֶרֶךְ אַפַּיִם). ◄ It was when Moses heard this measure of God's mercy that he fell upon his face in deference before God, thereby indicating the centrality of this trait.

In this book, we will see numerous examples of how writing verses in a figurate number form (based on the number of letters or words they possess) reveals intriguing findings that shed new light on many teachings by the sages.

The extension of the idea that square numbers represent the Kabbalistic notion of inter-inclusion is that every figurate number represents a concept in Torah. Without getting into the details, each of the three basic figurate forms—the triangle, square, and the interface—represent a certain form of relationship. In particular, triangular numbers represent a cause-effect (or evolutionary) relationship, the square numbers represent inter-inclusion as we saw, also known as an enclothement relationship (e.g., the soul

▶ Since this is the square of 13, it has a central-most element, the letter ח, whose value is 8. When we add it to the value of the corners, we get 360. Since the sum of 5 letters gave us this sum, their average value is thus 72, alluding to the filling of *Havayah* that equals 72 and is considered the highest of four essential fillings (see in length in *What You Need to Know About Kabbalah*, p. 143).

Mathematically speaking, the central-most element is the 85th letter. 85 is the interface number of 7, i.e. $85 = 7^2 + 6^2$. This follows the general relationship that the middle element of every odd square is the interface number of the square's root midpoint (for more about midpoints, see chapter 2, pp. 26ff.).

is enclothed in the body), and the interface numbers represent an inspirational relationship, which can also be termed as action-at-a-distance or non-locality.[1]

Systems of *Gematria*

Traditionally, there are a number of different mappings of letters to values used in analyzing the Torah's mathematical dimension. The most ubiquitous is called normative *gematria*, in which the letters take the values of the numbers, 1 through 9, 10 through 90, and 100 through 400. Among the other mappings used, we find the ordinal value, the reduced value, and the final-reduced value mappings. These are discussed in some detail in chapter 4 of the present volume.

The Golden Section

Though practically, every significant mathematical phenomenon can be useful in studying the Torah's quantitative dimension, one of the most beautiful and important is that of the Golden Section. The Golden Section was well known already in ancient times, and figures prominently in hundreds of places in the Torah.

The Golden Section spans a variety of mathematical branches. In a sense, it serves as a powerful bridge between continuous and discrete mathematics. In the realm of discrete mathematics, the Golden Section is related to additive series—series of numbers in which each number is the sum of the two previous numbers. The best known of the additive series is the Fibonacci series: 1, 1, 2, 3, 5, 8, 13, 21, 34, etc. In the Torah, this series is known as the Love series of numbers. We will see a few examples of how the Golden Section appears in the Torah in this present volume. However, we plan to publish a volume dedicated solely to the Golden Section phenomenon in the Torah.

---

1. For more on this correspondence, see *What You Need to Know About Kabbalah*, pp. 9–11.

Interpreting the findings

As mentioned, the authors of the traditional works incorporating *gematria*—the quantitative dimension of Torah—are considered to have possessed a minor form of prophecy known as *Ru'ach Hakodesh* (holy spirit). This figures most prominently in the qualitative meaning they ascribe to their findings. In a seminal discourse on the topic of allussions and *gematria* in Torah learning,[2] the Lubavitcher Rebbe writes that though one of the four methodologies for Torah study is *remez* (allusions), which includes the topic of *gematria*, not everyone is capable of properly using this method, as it requires the ability to properly interpet the meaning of the finding. Sometimes though, as mentioned above, it is the importance or complexity of the mathematics that warrants considering a certain finding and interpreting it.

For this reason, in the present volume, the mathematics and their meaning are interwoven throughout the text. In all, many of the techniques and phenomena described in this introduction are employed, but most importantly, every mathematical result is given due consideration from a traditional Torah point-of-view. The reader will find that many times the mathematics guides us to a deeper and more meaningful appreciation of the Torah's meaning.

Detailed overview of this book

The present work is a small sampling of Harav Ginsburgh's mathematical findings in Torah and their interpretations. All the findings center on the Torah's first verse, "In the beginning, God created the heavens and the earth," arguably one of the most important verses in the entire Torah, as the sages declare that, "All follows the inception." The findings and their interpretation have been presented in a way that will allow the reader to follow some of the main methodological points and techniques Harav Ginsburgh

2. *Likutei Sichot*, v. 26, pp. 204-208.

uses when analyzing the Torah mathematically. Thus, chapter 1 begins by looking at the verse's structure. The first quantitative question that should be asked about any verse is how many words and how many letters does it possess? The first chapter continues by looking at other central verses in the Book of Genesis and elsewhere in the Pentateuch that have an identical structure in terms of words and letters. Bringing these verses together sends the reader on an exploration of some of the most fundamental concepts in Torah related to the Working of Creation and the Working of the Chariot and their relationship with the foundations of the Oral Torah, the Mishnah and the Talmud. The first verse's structure then leads us into other related discussions regarding its form.

Chapter 2 introduces the concept of figurate numbers more formally, focusing of course on the triangular numbers, the form into which the Torah's first verse can be molded. A semi-rigorous discussion of the triangular numbers ensues, leading us to consider how the first verse actually exhibits all the triangles from 1 to 28. We further see how the entire verse's *gematria* value is a triangular number, connecting us with "wisdom" both as a Hebrew word and as a concept. This discussion of the triangular numbers' properties leads us to consider what are called midpoint series, an infinite family of series of integers, each beginning with a unique even number and developing into an infinite number of odd numbers.

Chapter 3 begins with a more conceptual Torah discussion of the concept of the midpoint and how it is used in different fields of Torah scholarship. It also reveals the essential connection between the Torah's first *parashah*, *Bereisheet*, and its second *parashah*, *Noach*. By looking at the midpoint series involving the middle word of the Torah's first verse, we come to consider verses that reveal this particular property through the juxtaposition of this word and the midpoint series' source. The first such verse is discussed as well as further properties of the midpoint series connected with square numbers. The chapter ends with a connection between midpoint series and the fine-structure constant in physics.

We open chapter 4 with an overview of how certain numbers are found to be linked ubiquitously throughout the Torah text. One such pair—23 and 37—is observed beautifully in the Torah's first verse. From here we move to a discussion of the various systems of *gematria*, different schemes for correlating the letters of the Hebrew alphabet with numbers. Chapter 5 continues the discussion in chapter 4 and explains the spiritual relevance and meaning of these different systems of *gematria*.

Chapter 6 focuses on the number 37—one of the central primes found in the Torah's first verse—and its word counterparts, "Abel" and "vanity." This leads into a discussion of how Abel was replaced with Seth and the verse describing this event. Considerations of the meaning of Seth's birth and of mankind's first family ensue, leading us to an exploration of more *gematria* systems and how numbers point our way in understanding the Torah's esoteric dimension.

With chapter 7, we take a step back to consider, much as science does today, what numbers are at the core of creation. Developing the case for 3 and 4 and their interrelated occurrences in Torah, we look at how these all allude back to creation and the Torah's first verse. This in turn offers us a chance to follow some algebraic reasoning and development of mathematical concepts as they service the study of Torah mathematics. The chapter concludes with a discussion of one of Rabbi Avraham Abulafia's most interesting mathematical findings relating to the Torah's first word and the number 913.

Chapter 8 revisits the question of what numbers lie at the root of creation, and illustrates how the Torah's first verse points us in the direction of the number 6. Additional findings that relate the number 6 to the Torah's first verse are discussed, findings that illustrate how positional values of letters in a verse can be used to analyze the Torah's quantitative dimension.

The final chapter, 9, considers permutations of a word, specifically analyzing 1 of the 720 possible permutations of the Torah's first word, "In the beginning," whose value is 913. Six different meanings of this permutation are explored in depth, both mathematically and from a Torah perspective. Finally all six meanings are tied back to fire, of which the sages say there are 6 types.

In all, there are a few dozen mathematical topics discussed in this book and a similar number of mathematical techniques used in Torah analysis are illustrated and explained. The book's flow is associative in nature, and this is an important point: the Torah's mathematical dimension is revealed not through an exhaustive process of research, but rather through higher inspiration that leads the learner form one finding to the next. With this book we are privy in part to watch the flow of Harav Ginsburgh's thought as it advanced from point to point ultimately creating a beautiful mosaic of numbers and linguistic meaning that continuously pivots around the central theme—the Torah's first verse and specifically, its first word.

Conventions

We are happy to utilize in this volume a new format for presenting Rabbi Ginsburgh's teachings. A word about the history of this present work is due. About 5 years ago, in the course of a week, Rabbi Ginsburgh wrote the text of this book—including, of course, all the mathematics, many of it incorporating new phenomena that came up during the course of writing. He then sent it to the editor, who proceeded to prepare it for publication by incorporating many of the very lengthy footnotes (at times running a page or two in length) back into the main text, as much as possible. However, even after this first edit was finished, it was clear that many of the footnotes were still far too long and though their contents were connected with the main text, they could certainly stand as a topic in their own right. About the same time as the first edit was complete, the editorial staff at Gal Einai began using Adobe Indesign, opening up new possibilities in laying out text. Based on inspiration received from two of our Russian-language editors, experimentation on a new format for this very mathematical book began. Eventually, we decided on the present layout. The most important change for the reader is that the lengthy footnotes have all been moved into side-boxes of text that are linked back to the main text with small yellowish triangles. The triangles act like arrows pointing the reader from within the main text to the appropriate side-box. The text in each side-box can be read as a self-contained topic. The topic heading appears in bold within the text itself. Employing side-text boxes for the longer topics included in the original footnotes has allowed us to utilize footnotes for citing sources or references to further reading only. We have opted not to include separate footnotes for citing the sources of the material covered in the side-boxes. Instead, all sources are cited inline, albeit in parentheses and a smaller font.

As in all our previous books, we adhere mostly to the guidelines set out in the Encyclopedia Judaica for transliterating Hebrew words in English. When transliterating God's essential Name, the Tetragrammaton, into English we use the accepted pronunciation, *Havayah*, which is actually a permutation of the four original letters. When transliterating other sanctified Names of God, we use the alternate, traditional pronunciations.

When providing the Hebrew original of verses or terms, we have vowelized the letters. Many times throughout the text, we refer to the value of a Hebrew word or phrase. In these cases, the Hebrew text—provided for reference—is printed in bold.

IN THE BEGINNING

The Basic Structure

The first verse of the Torah reads,

In the beginning God created the heavens and the earth

And, in the original Hebrew,

בְּרֵאשִׁית בָּרָא אֱלֹהִים אֵת הַשָּׁמַיִם וְאֵת הָאָרֶץ

The very first numerical analysis that should be done when meditating on any verse in the Bible is on its basic structure. The first quantitative questions we ask are: How many letters in the verse? How many words? And, revealing more structure, how are these words divided into phrases?

In the Torah's first verse, we find that the structure is that of 7 words and 28 letters divided into two phrases. The ratio of words to letters is 1:4, considered the ideal such ratio in the Torah because this is the number of letters in the Torah's most important word, *Havayah* (יהוה), God's essential Name.

The most important word in the Torah, God's essential Name *Havayah* (יהוה), comprises 4 letters▸, for which reason it is called the Tetragrammaton, literally "the four-letter Name" (שֵׁם בֶּן ד). This is the idiom that the sages use to refer to the Name *Havayah*, implying that the fact that the holiest of God's Names possesses four letters is essential to its meaning.▾▾

▸ Most roots in **the Hebrew language**, the language of creation, possess 3 letters. This means that in general for a word to have 4 letters it must comprise a Hebrew root together with an additional letter, either before the root letters (a prefix letter), after them (a suffix letter), or in the middle of the root. In the case of verbs, the future tense is most often formed by adding a prefix letter in place of the letter *yud* of the form יִפְעַל; the present tense, by adding a middle letter in place of the letter *vav* of the form פּוֹעֵל; and, the past tense, by adding a suffix letter in place of the letter *hei* of the form פָּעֲלָה. Note that the 3 letters *yud*, *vav*, and *hei*—added to the past, present, and future forms of a verb—are the letters that comprise God's essential Name, *Havayah* spelled *yud-hei-vav-hei*.

letter of *Havayah*	tense	form	meaning in context of *Havayah*
yud (י)	future	יִפְעַל	he will be
hei (ה)	past	פָּעֲלָה	she was
vav (ו)	present	פּוֹעֵל	he is

In Kabbalah it is explained that the letters *yud* and *vav* in *Havayah* are relatively masculine letters while the *hei* is relatively feminine. In addition, the final *hei* of the Name reflects the first *hei*, meaning that the origin of the female past (she was) unites with the male future (he will be) in order to reveal, to give birth to, the male present (he is). This idea finds expression in the instruction of the sages that when contemplating the Name *Havayah* we are to have in mind that its letters form the three words: "[He] was" (הָיָה), "[He] is" (הֹוֶה), "[He] will be" (יִהְיֶה), indicating that God is eternal and in essence above time. Indeed, time is God's first creation, as alluded to in the Torah's first word, "In the beginning."

▸▸ In Kabbalah, the four letters of *Havayah* correspond to the four general stages of the creative process and to the **four Worlds of creation**:

letter of *Havayah*	creative process	World
yud	contraction	Atzilut (emanation)
hei	expansion	Beri'ah (creation)
vav	extension	Yetzirah (formation)
hei	2nd expansion	Assiyah (action)

The sages consider the entire Torah to be one great Name of God. The Ga'on of Vilna explains that each word of the Torah can be categorized into one of the four categories of Divine Names (*Commentary to Sefer Yetzirah*). These four categories themselves correspond to the 4 letters of God's essential Name *Havayah* (See *What You Need to Know About Kabbalah*, pp. 177ff.).

In his Guide to the Perplexed (1:69 and 1:72), Maimonides explains in length the philosophical significance of the number 4 in the Torah.

And so, even though the Name *Havayah* does not appear explicitly in the first account of creation in the Torah, it is alluded to in the structure of the first verse.

We noted that the first verse has 28 letters. 28 is the *gematria* of the word for "power" or "force" (כֹּחַ), alluding to the verse that *Rashi* quotes in his commentary to the first verse of the Torah, "The power of His actions He told His people, to give them an inheritance of nations"[1] (כֹּחַ מַעֲשָׂיו הִגִּיד לְעַמּוֹ לָתֵת לָהֶם נַחֲלַת גּוֹיִם).

The cantillation marks, which also serve as punctuation, divide the first verse of the Torah into two major phrases (as is the case for almost every verse in the Torah). Each phrase contains 14 letters. 14 is the *gematria* of "hand" (יָד). Thus, the Divine power (the total 28 letters of the first verse) of creation manifests in the two "hands" of the Creator, as it were. The 14 letters of "In the beginning God created" (בְּרֵאשִׁית בָּרָא אֱ-לֹהִים) correspond to the right hand and the 14 letters of "the heavens and the earth" (אֵת הַשָּׁמַיִם וְאֵת הָאָרֶץ) correspond to the left hand. This division between the right and left aspects of creation reflects the difference between creation as an explicit act by the Almighty, as described by the first 14 letters, "In the beginning God created," and creation as an anonymous act, without author as it were, as described by the second 14 letters, "the heavens and the earth."

Of course, what may seem anonymous is not, and as the prophet Isaiah[2] states, God created the heavens with His right hand and the earth with His left hand. Thus, the action of God's left hand as it were, itself divides into a relative right (heavens) and a relative left (earth)—relative to the left as a whole. This is an example of the Kabbalistic principle of inter-inclusion (הִתְכַּלְלוּת); in this case left and right aspects are inter-included within the left hand.

We will now look at other verses in the Torah that have the same structure as the first verse, 7 words and 28 letters. ◄

▶ **Comparing verses** that have an identical structure (number of words and letters) is a technique employed by the sages of earlier generations, even in determining issues relating to liturgy. See the *Ba'al Hatureem* on this verse (specifically, the Jerusalem: Feldheim, 5756 edition edited by Jacob Reinitz, p. 5-6), and the *Arba'ah Tureem* to *Orach Chayim* 56.

---

1. Psalms 111:6.
2. Isaiah 48:13.

Genesis and the Ten Commandments

Another most central verse in the Torah that possesses the same basic structure of 7 words and 28 letters precedes the Ten Commandments,[3]

God spoke all these things▶ saying.▶▶

And, in the original Hebrew,

וַיְדַבֵּר אֱ־לֹהִים אֵת כָּל הַדְּבָרִים הָאֵלֶּה לֵאמֹר

There are many phenomena, both quantitative and qualitative that connect these two verses, and their identical structure is like a marker guiding us to look at them together. Of course, the general point we are making is that whenever two verses have the same structure, we are led to look at their relationship. Let us see some of these phenomena.

The first thing to note is that the two verses share a common prime factor: 37.

The *gematria* of the first verse of the Torah is 2701 = 37 · 73 (= △73, read: the triangle of 73). The *gematria* of the verse preceding the Ten Commandments is 1332 = 37 · 36 (the double triangle or diamond▶▶▶ of 36). But, now note that the difference between their values is not just a multiple of 37, it is 37^2!

37^2 = 1369, a number that in itself makes meaningful appearances both in relation to the Ten Commandments and in relation to the beginning of Genesis. Regarding the first, 1369 is the number of letters in the Torah section (פָּרָשָׁה) that precedes the verse before the Ten Commandments. This section describes the preparation for the receiving of the Torah.[4] Regarding the second, 1369 is the value of the second half of the Torah's second verse, "...And the spirit of God hovered over the face of the waters" (וְרוּחַ אֱ־לֹהִים מְרַחֶפֶת עַל פְּנֵי הַמָּיִם).

Returning to 37, it is a prime factor of the last word of our verse

---

3. Exodus 20:1. See *Rashi's Sefer Hapardes* (Ehrenreich edition, p. 92), which explains the structural and essential connections between the two verses.
4. Exodus 19:1-25.

▶ All things were created as a reflection of their **inner essence**, captured by the Hebrew word that is their name.

Indeed, in Hebrew the word for "thing" (דָּבָר) also means "word," for example in the verse (1 Kings 2:14), "I would like a word with you" (דָּבָר לִי אֵלֶיךָ). The names of all things originates in the Ten Utterances, the ten times God is described as speaking in the account of creation.

▶▶ The most often-repeated verse in the Pentateuch is, "God spoke to Moses, saying" (וַיְדַבֵּר יהוה אֶל מֹשֶׁה לֵּאמֹר). Normally, **the word "saying"** (לֵאמֹר) means that Moses was to publicize the words spoken to him by God to the entire people.

Of course, this meaning is not easily carried over to the verse preceding the Ten Commandments, since God was speaking to the entire Jewish people. Therefore, the Chassidic masters interpret it to mean that the words of the Ten Commandments (and of the Torah in general) will continue to forever reverberate in the mouths and souls of the Jewish people.

▶▶▶ The **diamond** of n is a figurate number designated by the symbol ◆n and defined as 2 times the sum of integers from 1 to n. Using mathematical notation we write,

$$◆n = 2△n$$

The first few diamond numbers are thus,

2, 6, 12, 20, 30, 42, ...

Diamond numbers are easily pictured. Here are the first few forms of the diamond numbers,

▶ In the verse: "God spoke all these things, saying" the root that also means "thing" ד.ב.ר appears twice, in the words "spoke" (וַיְדַבֵּר) and "things" (הַדְּבָרִים). The *gematria* of these two words together is 483 = 21 · 23. 23 is considered the companion number of 37 in Kabbalah.

23 and 37 are considered idiomatic, or companion numbers in Kabbalah, meaning that they tend to appear together when we explore the Torah's quantitative aspect.

▶▶ The first few numbers in the series of **double square numbers** are: 2, 8, 18, 32, 50, 72, …

These numbers are defined by the function $f[n] = 2n^2$.

Double squares play a prominent role in Kabbalah, the most famous of them being the 32 Pathways of Wisdom, the 50 Gates of Understanding, and the 72 Bridges of Knowledge.

In science, they figure prominently in chemistry where they describe the number of possible electron orbitals, one of the underlying physical properties that gives the Periodic Table of Elements its structure.

Double square numbers can be drawn in a distinctive form. Here are the first few in their figurative form,

from Genesis, "the earth" (הָאָרֶץ), 296 = 8 · 37 and a prime factor of the first word of our verse from Exodus, "spoke" (וַיְדַבֵּר), 222 = 6 · 37◀. Obviously the sum of these two words, 518, is also a multiple of 37.

Switching the order, the first word of our verse from Genesis, "In the beginning" (בְּרֵאשִׁית), added to the last word of our verse from Exodus, "saying" (לֵאמֹר), gives 1184, also a multiple of 37, 1184 = 32 · 37. Adding all four words (first and last from both verses) gives us, 518 + 1184 = 1702 (a permutation of the digits of 2701) = 2 · 23 · 37. One more word is itself a multiple of 37, the sixth word of the first verse, "and" (וְאֵת), 407 = 11 · 37. The remaining 9 words of both verses equal 1924 = 52 · 37 = 74 · 26, where 74 is the *gematria* of the word "witness" (עֵד) and 26 is the *gematria* of the Name *Havayah*, alluding to the verse, "*Havayah* is witness in you" (עֵד יְהוָה בָּכֶם).[5]

The Ten Commandments, as they appear in the Book of Exodus (*parashat* Yitro) contain 620 letters. The Ten Commandments appear a second time in the Book of Deuteronomy where they include an additional 88 letters. These additional 88 letters allude to the word "creek" (נַחַל). The creek is not mentioned in Exodus, but is described by Moses as flowing down Mt. Sinai in Deuteronomy.[6] Now, 620 is the *gematria* of the *sefirah* of "crown" (כֶּתֶר), the highest of the ten *sefirot* (the Divine emanations through which God created the world), which corresponds to the super-conscious realm of the soul. Thus, the verse that precedes and introduces the Ten Commandments (in Exodus) is regarded as the crown above the crown. Adding the 28 letters of the verse introducing the Ten commandments to the 620 letters of the Ten Commandments gives 648 letters. $648 = 2 · 18^2$ (also called the double-square◀◀ of 18) where 18 is the value of "alive" (חַי), the well-known word that is pronounced *chai*! The difference between 620 (the number of letters in the Ten Commandments) and 28 (the number of letters in the verse preceding them) is 592, one of whose prime factors is 37, specifically, 592 = 16 · 37. We will

5. 1 Samuel 12:5.
6. Deuteronomy 9:21.

presently see the importance of the number 37 in relation to the two verses that possess the identical structure of 7 words and 28 letters.

From the introductory verse of the Ten Commandments, the sages learn that all of the Ten Commandments were first spoken by God as a single seminal word.[7] The sages[8] identify the Torah's first verse as the (implicit) first of Ten Utterances, with which God created reality. Similarly, the sages state that this first utterance is seminal,[9] and with it all the particulars of creation were put in place. Like the Ten Commandments, the Ten Utterances also correspond to the ten *sefirot* and therefore the first verse of Genesis is regarded as the crown of creation just as the verse preceding the Ten Commandments is the crown of the Giving of the Torah.

Another fact underlying the connection between these two verses is the Name of God, *Elokim*, found in both. Obviously, the Name *Elokim* is common in the Torah, but these two verses stand out. The first verse of Genesis, because it contains the first time *Elokim* is mentioned. The appearance of *Elokim* in the verse preceding the Ten Commandments stands out because of the very rare idiom "God [*Elokim*] spoke" (וַיְדַבֵּר אֱלֹהִים) it opens with. While "God [*Havayah*] spoke" (וַיְדַבֵּר יהוה) is common in the Torah, the phrase "*Elokim* spoke" is very rare and in fact appears only 3 times in the entire Bible. The appearance here, where it crowns the Ten Commandments, is its third and final. Let us diverge for a moment to look at these three appearances.▸

The phrase "God [*Elokim*] spoke" first appears in the verse, "*Elokim* spoke to Noah saying"[10]; this is also the first time that the word "spoke" (וַיְדַבֵּר) appears in the Torah. Its second appearance is in the verse, "*Elokim* spoke to Moses and said to him, 'I am *Havayah*.'"[11]

---

7. See *Rashi* on the verse (Exodus 20:1) based on *Mechilta Yitro parashah* 4.
8. *Rosh Hashanah* 32a.
9. Rabbi Nechemiah in *Midrash Bereisheet Rabbah* 12:4. For more on this topic, see our forthcoming volume on Evolution.
10. Genesis 8:15.
11. Exodus 6:2.

▸ Referring to the speaker, *Elokim*, these three verses follow the spiritual order of Divine service and transformation taught by the Ba'al Shem Tov: **submission, separation, and sweetening** (as explained at length elsewhere, see in particular *Transforming Darkness into Light*).

The verse addressing Noah represents a state of submission. The nature of Noah's relationship with God was submissive, in that he could do no more than accept the world's impending destruction (as opposed to Abraham, the first Jew, who when informed of God's plan to destroy Sodom and Gemorrah, did all he could to argue against the decree).

The first verse addressing Moses (Exodus 6:2) serves as the opening to the Ten Plagues, whose essence was to separate between the Jewish people and the Egyptians in preparation for the ultimate separation between them, which occurred with the Exodus and the Splitting of the Red Sea.

Finally, the verse preceding the Ten Commandments represents the ultimate state of sweetening attained with the Giving of the Torah at Mt. Sinai.

5

▶ The sages (*Shabbat* 88b) use the phrase "from the mouth of **might**" (מִפִּי הַגְּבוּרָה) to describe the source from where we heard the Ten Commandments. Indeed, among God's different Names, *Elokim* is the one that corresponds to the *sefirah* of might (See *What You Need to Know About Kabbalah*, pp. 152-3).

The appearance of the Name associated with might in this introductory verse refers to the might with which God concentrated His infinite light into the finite number of letters of the Ten Commandments (and the Torah in general).

These first two instances have something interesting in common. The first is God's command to Noah to exit the ark in which he had been shipbound for a year. The second verse is spoken by God to Moses, commanding him to take the Jewish people out of their bondage in Egypt. The third instance is also connected to the Exodus from Egypt, since it introduces the Ten Commandments that begin with "I am *Havayah* your God who has taken you out of the Land of Egypt from a house of slaves."[12] So even though we usually associate God's Name *Elokim* with limitations (such as the limited laws of nature, which it represents) at a higher level it also represents the power to emerge free of contraction. This is explained in Chassidic philosophy as *Elokim* functioning not to limit, but to concentrate. ◀ Thus, in the Ten Commandments God limited, in the sense of concentrated, His essence as it were, into the words spoken at Mt. Sinai. The archetypal example of this concept is found in regard to the Ark of the Covenant, about which the sages say, "He concentrated the Divine Presence in between the carrying rods of the ark."[13] The concentration of the Divine Presence promotes the Almighty to speak directly through a person, as He did through Moses, but for this to occur, one must first break the bonds of Egypt, i.e., the Egyptian conceptual scheme that is based on a sense of being separate from the Almighty, causing the Name *Elokim* (אֱ-לֹהִים) to reveal that it is also equal to "the vessel of *Havayah*" (כְּלִי י-הוה). A similar state is described by King David in Psalms, "I [God] have said that you [mankind] are *Elokim*."[14]

Adding the *gematria* of the first verse of the Torah, 2701, to the *gematria* of these three verses results in 5776, a perfect square (76^2), indicating that these four verses do indeed form a related set. This motivates us to look at their structure a bit more. Together, all three verses containing this rare phrase, "*Elokim* spoke" have 77 letters.

---

12. Exodus 20:2.
13. *Tanchuma Vayakhel* 7 and elsewhere, that the Divine Presence was concentrated in the Ark of the Covenant.
14. Psalms 82:6.

Together with the 28 letters of the Torah's first verse, the total comes to 105 letters, which allows us to arrange all 4 verses in the form of a triangle (specifically the triangle of 14, a topic we will return to cover in greater detail in the next chapter), like so,

```
                              ב
                           ר     א
                        ש     י     ת
                     ב     א     ר     א
                  ל     ה     י     ם     א
               ת     ה     ש     מ     י     ם
            ו     א     ת     ה     א     ר     ץ
         ו     י     ד     ב     ר     א     ל     ה
      מ     א     ל     נ     ח     ל     א     ם     י
   י     ה     ל     א     ר     ב     ד     י     ו     ר
ר     מ     א     י     ו     ה     ש     מ     ל     א     ם
א     ל     י     ו     א     נ     י     י     ה     ו     ה     ו
ל     כ     ת     א     ם     י     ה     ל     א     ר     ב     ד     י
ר     מ     א     ל     ה     ל     א     ה     ם     י     ר     ב     ד     ה
```

Note that in this triangle, the first 7 rows (the triangle of 7) contain only the 28 letters of the first verse. The bottom 7 rows contain the other 3 verses. Thus the first verse alone makes up the triangle's upper half.

Returning to the first verse in the Torah, its *gematria* is 2701, or the triangle of 73 (the sum of integers from 1 to 73, again, a topic we will explore a great deal more in chapter 2). For now, let us note that 73 is the *gematria* of "wisdom" (חָכְמָה), the name of the *sefirah* following crown. Adding 620, the value of "crown" (כֶּתֶר), to 2701 gives us 3321, also a triangular number, this time the triangle of 81 (again, the sum of integers from 1 to 81). But, this connects us back to the Ten Commandments, because 81 is the *gematria* of "I" (אָנֹכִי), ▸ the first word in the Ten Commandments. In fact, 3321 is also a multiple of 81. Specifically, 3321 = 81 · 41, where 41 is the number of letters in the first verse of the Ten Commandments, "I am *Havayah* your God who has taken you out of the land of Egypt from a house of slaves". 41 is also the value of the first and last letters in this verse, (א and ם).

▸ The sages explain (*Shabbat* 105a) that the four letters making up the first word of the Ten Commandments, "I" (אָנֹכִי) are initials of the phrase, "I have written and given myself" (אֲנָא נַפְשִׁי כְּתָבִית יְהָבִית), suggesting as the *Zohar* (III, 73a) states that **the Almighty and the Torah are one.**

▶ The Ba'al Shem Tov teaches us that the essence of our Divine service is to unite **Worlds, Souls, and Divinity** (see the introduction to The Hebrew Letters). Though the Divine is constantly sustaining all Worlds, in what is referred to as the power of the Maker in that which is made (כֹּחַ הַפּוֹעֵל בְּנִפְעָל), without the presence and consciousness of Souls, only a minute reflection of this power can be revealed.

The manifestation of this minute reflection of Divinity in Worlds is the secret of the Name *Elokim* (אֱ-לֹהִים) whose numerical value is the same as the Hebrew word for "nature" (הַטֶּבַע) and is the basis of the mystery of the Working of Creation. But through the means of Souls the full essence of Divinity—the secret of God's essential Name, *Havayah* (the basis of the mystery of the Working of the Divine Chariot)—that continuously sustains reality can become revealed.

Furthermore, the 41st prime (when including 1 as the first prime), is 173, which is the value of the first 3 words of the first commandment, "I am *Havayah* your God…" (אָנֹכִי יְהוָה אֱ-לֹהֶיךָ).

Two Great Mysteries

The sages[15] speak of two great[16] mysteries, the Working of Creation (מַעֲשֵׂה בְּרֵאשִׁית) and the Working of the Divine Chariot (מַעֲשֵׂה מֶרְכָּבָה).

▶▶ The mysteries of the Torah are also known as the *sod*, the secret of the Torah. They are symbolized by wine, of which it is said that, "when wine enters, secret comes out" (Eiruvin 65a). The *gematria* of wine (יַיִן), 70, is equal to that of secret (סוֹד).

The mysteries of the mysteries are symbolized by pure olive oil, the source of radiance. **Oil** (שֶׁמֶן), 390, is a multiple of 13 while **wine** (יַיִן), 70, is a multiple of 7. 7 and 13 are companion numbers (like 23 and 37) that appear together throughout the Torah. The number 7 alludes to the 7 days of creation while the number 13 alludes to the 13 Attributes of Divine Mercy, revealed in reality through the conduit of the Divine soul of Israel, God's Divine Chariot.

The Working of Creation is the mystery of creation *ex-nihilo*, a continuous act of the Creator. The Working of the Divine Chariot is the mystery of how God reveals Himself within reality through the means of the Divine soul of Israel.▲ The Working of Creation is "the mystery of the Torah" (רָזִין דְּאוֹרָיְתָא) while the Working of the Divine Chariot is "the mystery of mysteries of the Torah" (רָזִין דְּרָזִין דְּאוֹרָיְתָא).◀◀

These two great mysteries make up the top two levels of the structure of the entire Torah, which in its widest sense possesses five levels, three revealed and two concealed. These are three revealed:

1. Scriptures (the Bible)
2. Mishnah
3. Gemara

and two concealed:

4. Working of Creation
5. Working of the Divine Chariot ◀◀◀

▶▶▶ Incredibly, the account of the Working of the Divine Chariot (Ezekiel ch. 1) contains 382 words, which together with the 469 words in the account of the Working of Creation (Genesis 1:1 through 2:3) totals **851 words**. But, 851 is the product of 23 and 37, the values of the "living one" (חַיָּה) and "singular one" (יְחִידָה), respectively!

---

15. *Chagigah*, chapter 2.
16. See *Sukah* 28a and *Bava Batra* 134a.

These 5 levels of the Torah correspond to the five levels of the soul, which is also divided into 3 (lower) revealed and conscious levels and 2 (higher) concealed and super-conscious levels.

The correspondence between the two is thus,

part of Torah	part of soul
Working of the Divine Chariot	the singular-one (יְחִידָה)
Working of Creation	the living-one (חַיָּה)
Gemara	inner-mind or breath-of-life (נְשָׁמָה)
Mishnah	spirit (רוּחַ)
Scriptures▶	life-force (נֶפֶשׁ)

Thus, the Working of Creation is the living-one (חַיָּה)▶▶ of the Torah (the Divine source of all life in creation, which radiates life-force to the lowest spiritual level of life-force (נֶפֶשׁ), reflected by the fact that the Working of Creation appears at the beginning of the Scriptures) and the Working of the Divine Chariot is the single-one (יְחִידָה) of the Torah (the revelation that all is in essence one—"*Havayah* is One").▶▶▶ In passing, note that the value of the soul's two super-conscious levels, the living one (חַיָּה) and the singular one (יְחִידָה) is 23 and 37, respectively, a central example of 23 and 37 as companion numbers.

The Talmud states this relationship using the Talmudic debates of the two great sages Abayei and Rava as representative of the entire revealed Torah. The mystery of the Working of the Divine Chariot are then referred to as "a great thing" (דָּבָר גָּדוֹל),▶▶▶▶ relative to the debates of Abayei and Rava, described as "a small thing" (דָּבָר קָטָן). The mystery of the Working of Creation is an intermediate level between the Working of the Divine Chariot and the debates of Abayei and Rava, a small thing relative to the former and a great thing relative to the latter. Thus, it is the Working of Creation—the secrets of nature—that unify the two extremes of great and small. Indeed, the most powerful idea in studying the secrets of creation is that the macrocosm and the microcosm are identical in structure.

There is a general principle in Torah that in order to act as a unifying

▶ The numerical value of "**Scriptures, Mishnah, Gemara, Working of Creation, Working of the Chariot**" (מִקְרָא מִשְׁנָה גְּמָרָא מַעֲשֶׂה בְרֵאשִׁית מַעֲשֶׂה מֶרְכָּבָה) is 2990, which obviously divides by 5, so their average value is 598 = 26 · 23, where 26 is the value of God's essential Name, *Havayah*, and 23 the value of "the living one" (חַיָּה).

▶▶ That the Working of Creation corresponds to the fourth level of the soul, **the "living one"** is beautifully alluded to in the actual text in Genesis that describes the Working of Creation. The first account of creation contains 469 words. Its exact middle is the 235th word appearing in Genesis 1:20, which is none other than "living one" (חַיָּה). This is also the first appearance of this word in the Torah!

▶▶▶ The sages explain that during the Giving of the Ten Commandments at Mt. Sinai, every Jew experienced the Working of the Divine Chariot. The account of the Working of the Divine Chariot (Ezekiel ch. 1) contains 382 words. Adding 238, a permutation of the same digits, gives 620, **the number of letters in the Ten Commandments!**

▶▶▶▶ The *gematria* of "**a great thing—** the Working of the Divine Chariot" (דָּבָר גָּדוֹל מַעֲשֵׂה מֶרְכָּבָה) is 931, also the value of the three dimensions of reality as defined in *Sefer Yetzirah*, "world, year, soul" (עוֹלָם שָׁנָה נֶפֶשׁ). While the Torah was being given on Mt. Sinai, all three dimensions united into one. This is the secret of the "smoke" that the Torah describes rising from Mt. Sinai. The initial letters of "world, year, soul" (עוֹלָם שָׁנָה נֶפֶשׁ) spell "smoke" (עָשָׁן).

intermediate between two things or concepts, the intermediate must actually be higher than both, as it must within itself already contain aspects of both extremes it is uniting. So it is in regard to the Working of Creation which here connects the levels of Gemara—the debates of Abayei and Rava—and the Working of the Chariot. This is the case because like the Torah, the levels of the soul are not distinct entities. They are in effect a single continuum broken up by functionality or concentration. However, much like a holographic crystal, every part contains the whole, thus parts that are in effect intermediaries can be actually seen in practice to contain the parts bordering with them. In our case, Abayei's name (אַבַּיֵי) actually equals "the living one" (חַיָּה), whereas Rava's name (רָבָא) even though it is composed of the same letters as the first verb in the Torah, "created" (בָּרָא), because of the particular order of the letters, actually refers to the Working of the Chariot. This can be seen by analyzing the letters of Rava in comparison with those of "chariot" and of "creation." The 3-letter root of "chariot" is רכב, the 3-letter root of "creation" is ברא. The two letters that are shared are *reish* (ר) and *bet* (ב), except that in the chariot, the *reish* comes before the *bet*, and in "creation" their order is reversed. Indeed, the order רב, indicates greatness, just as in the sages' idiom, Working of the Chariot is described as "a great thing." The order בר, the order found in "creation," indicates manifestation, or extraction, the essence of the Working of Creation, which manifests the Creator's will into our reality.

What we have gained from this analysis is that within the Talmud, represented by Abayei and Rava, we find that Abayei corresponds to the Torah's level of "the living one" (חַיָּה) the Working of Creation, while Rava corresponds to the Torah's "singular one" (יְחִידָה). Indeed, we find that in the debates between these two great sages, the legal ruling almost always follows Rava.◄

The Working of Creation is described in the first account of creation in the Torah. The Working of the Divine Chariot ◄◄ is described in the first chapter of the book of Ezekiel.◄◄◄ In the Five Books of

▶ In the **debates** between these two great sages, the legal ruling follows Rava apart from 6 specific issues known by their acronym, יע"ל קג"ם (pronounced *ya'al kagam*) in which the ruling follows the opinion of Abayei, Rava's famous disputant (See *Rashi* to *Kidushin* 52a).

Incredibly the value of the acronym chosen to represent these 6 exceptional issues is 253, or 11 · 23, where 23 is the value of Abayei (אַבַּיֵי) and "living one" (חַיָּה).

▶▶ From the outset of creation, all of history is like a ride towards **the ultimate purpose of creation**, to make this world a dwelling place for God, a place where His very essence can become revealed to all. The ride is full of battles, between opposing forces of good and evil, light and darkness, for which reason a chariot throughout the Scripture is a vessel of war, used like a tank.

▶▶▶ Interestingly, the word "**chariot**" (מֶרְכָּבָה) does not appear in the Biblical account of Ezekiel's prophecy of the Divine Chariot. It is rather implied by the description of the four "living beings" (חַיּוֹת) and the spinning "wheels" (אוֹפַנִּים) that seem to be carrying them.

Moses,[17] the parallel to the first chapter of Ezekiel is the account of the Giving of the Torah to Israel,[18] culminating in the revelation of the Ten Commandments. ▶ For this reason on the holiday of *Shavuot*, the holiday commemorating the Giving of the Torah, we read the Torah portion culminating with the Ten Commandments and for the *haftarah*, the supplementary reading from the prophets, we read the vision of the Divine Chariot, the first chapter of Ezekiel.

The Patriarchs

"The fathers [i.e., the three Patriarchs, Abraham, Issac, and Jacob] are the Divine Chariot."[19] They were the selfless conduits that revealed Godliness in the world, upon whom God "rode,"▶▶ as it were.

The sages learn this from two verses in Genesis,[20] one referring to Abraham and the other to Jacob. The first verse[21] reads,

> He concluded speaking with him and God ascended from over Abraham

And in the original Hebrew,

וַיְכַל לְדַבֵּר אִתּוֹ וַיַּעַל אֱ-לֹהִים מֵעַל אַבְרָהָם

The second[22] verse reads,

> God ascended from over him [Jacob] in the place that He had spoken with him.▶▶▶

And, in Hebrew,

וַיַּעַל מֵעָלָיו אֱ-לֹהִים בַּמָּקוֹם אֲשֶׁר דִּבֶּר אִתּוֹ

---

17. "There is nothing in the Prophets and the Writings that is not alluded to in the Five Books of Moses" (see *Ta'anit* 9a).
18. It is explicitly stated (Psalms 68:18) that God appeared on His Divine Chariot at the Giving of the Torah to Israel.
19. *Bereisheet Rabbah* 82:6.
20. See *Rashi* on Genesis 17:22 and *Ramban* on Genesis 35:13.
21. Genesis 17:22.
22. Genesis 35:13.

▶ The value of the 6 words "Workings of Creation, living-one, Workings of the Divine Chariot, single-one" (מעשה בראשית חיה מעשה מרכבה יחידה) is 2070. Since 2070 is a multiple of 6, this means that the average value of these 6 words is 2070/6 = 345, the value of "Moses" (משה). This *gematria* alludes to **Moses' soul-root**, through which these two highest aspects of the Torah are revealed to us.

Indeed, the 3 letters of "Moses" (משה) have 6 unique permutations whose sum if therefore 2070. Note that 2070 is the diamond of 45 (see p. 3), where 45 is the value of "Adam" (אדם) whom Moses is considered to be the rectification of.

▶▶ Note that in the text it is explained that the level of the Divine Chariot as manifest in the Patriarchs and their Divine service is itself an intermediate level between creation and the Giving of the Torah to Israel, and so the fact that in the union of **Abraham and Sarah** she is greater than him in prophecy suggests that their connection is primarily in the Working of Creation, whereas the union of Jacob and his wives is strongly related to the Working of the Divine Chariot as manifest in the Giving of the Torah.

▶▶▶ The *gematria* of the first three words in this verse, "And God ascended from over him" (וַיַּעַל מֵעָלָיו אֱ-לֹהִים) is 358, the value of **Mashiach** (משיח).

The *gematria* of the remaining four words, "in the place that he had spoken with him" (בַּמָּקוֹם אֲשֶׁר דִּבֶּר אִתּוֹ) is 1302, or 7 · 186, where 186 is the value of "place" (מקום). The word "place" is a connotation for God, who is referred to as "the Place of the world" (מקומו של עולם). When the word "place" is written in full (מם קוף וו מם), the value of its letters equals 358, Mashiach!

Both of these verses contain 7 words and 28 letters! In each of these two verses the Name of God used is *Elokim*.

The secret of, "the fathers are the Divine Chariot" is an intermediate level of Divine revelation that lies between that of creation, where God's revelation is limited to the garb of nature and its wonders and that of the Giving of the Torah to Israel, in which God has revealed His very essence to Israel and the world. By definition, to function properly, every intermediate must possess two aspects or levels: one similar and connecting to each of the two extremes that the intermediate unifies. In our case, the chariot quality of Abraham connects to the Working of Creation, for God created the world with Abraham's soul-root.◄ The chariot quality of Jacob connects to the essence of the Working of the Divine Chariot (the essence of Divine revelation in the world), the Giving of the Torah to Israel, since Jacob corresponds to the pillar of Torah.◄◄

We can now understand why there is no explicit verse in the Torah demonstrating that Isaac is God's chariot, for he is the hidden force that unites the two sides of the revealed intermediate chariot quality that unites the two extremes of the Working of Creation and the Working of the Divine Chariot.

So far, we have 3 verses with the structure of 7 words and 28 letters that specifically relate to the Divine Chariot:

וַיְכַל לְדַבֵּר אִתּוֹ וַיַּעַל אֱ־לֹהִים מֵעַל אַבְרָהָם.
וַיַּעַל מֵעָלָיו אֱ־לֹהִים בַּמָּקוֹם אֲשֶׁר דִּבֶּר אִתּוֹ.
וַיְדַבֵּר אֱ־לֹהִים אֵת כָּל הַדְּבָרִים הָאֵלֶּה לֵאמֹר.

Their combined *gematria* is 4291, whose factorization is 7 and 613, specifically, 4291 = 7 · 613, where 613 is of course the number of commandments in the Torah.◄◄◄

When we add 4291, to the value of the Torah's first verse, our impetus for looking at verses with a 7, 28 structure, the total *gematria* becomes 6992 = 23 · 304. 23 is the value of "the living one" (חַיָּה) and 304 is the value of the two words, "He ascended… in the place"

▶ The first verse of the second account of creation is (Genesis 2:4), "These are the chronicles of the heavens and the earth when they were created." The letters of the word, "when they were created" (בְּהִבָּרְאָם) permute to spell "with Abraham" (בְּאַבְרָהָם).
Additionally, Abraham is the archetypal soul of loving-kindness (חֶסֶד) and loving-kindness too is described as **the building block of the world**, as revealed by the verse, "The world is built from loving-kindness" (Psalms 89:3).

▶▶ The sages state that, "The world stands on three things, on the Torah, on Divine service, and on acts of loving-kindness" (Avot 1:2). These **three pillars** correspond to the three Patriarchs: Abraham to acts of loving-kindness, Isaac to Divine service (sacrifices in the Temple service and prayer), and Jacob to Torah.

▶▶▶ Every one of the Torah's commandments is meant to aid us in **transforming ourselves** into a Divine chariot.

(וַיַּעַל...בַּמָּקוֹם), the two middle words of the verses of the Patriarchs and the first and fourth words of the verse of Jacob.

The *gematria* of the first words of the verses regarding Abraham and Jacob, "He concluded" and "[God] ascended" (וַיְכַל וַיַּעַל) is 182, the numerical value of Jacob (יַעֲקֹב).

The *gematria* of the first words of all four verses (בְּרֵאשִׁית וַיְכַל וַיַּעַל וַיְדַבֵּר) is 1317, the value of the four living-beings (חַיּוֹת) of the Divine Chariot, the lion, the ox, the eagle, and the man (אַרְיֵה שׁוֹר נֶשֶׁר אָדָם).

Drawing the verses

All four verses possess the same basic structure of 7 words and 28 letters. Together they possess 28 words (the number of letters in each verse, the triangle of 7, i.e., the sum of integers from 1 to 7) and 112 letters. 112 is the *gematria* of *Havayah Elokim* (יְהוָה אֱ־לֹהִים), whose numerical values are 26 and 86, respectively. This is called God's "full Name" (שֵׁם מָלֵא), the Name that expresses the absolute unification of the Working of Creation (*Elokim*) and the Working of the Divine Chariot (*Havayah*), the secret of these four verses!

Since all 4 verses together possess 28 words, we can order them in the form of a triangle of 7. Let us present them as such in the order that they appear in the Torah with the two intermediate verses related to the Patriarchs between the verse of creation and the verse of the Giving of the Torah,

בְּרֵאשִׁית

אֱלֹהִים בָּרָא

וְאֵת הַשָּׁמַיִם אֵת

אִתּוֹ לְדַבֵּר וַיְכַל הָאָרֶץ

וַיַּעַל אַבְרָהָם מֵעַל אֱלֹהִים וַיַּעַל

אִתּוֹ דִּבֶּר אֲשֶׁר בַּמָּקוֹם אֱלֹהִים מֵעָלָיו

לֵאמֹר הָאֵלֶּה הַדְּבָרִים כָּל אֵת אֱלֹהִים וַיְדַבֵּר

▶ When the Torah's first verse is depicted in the triangular form (see p. 27) the final row contains the two words "and the earth" (וְאֵת הָאָרֶץ) corresponding to the *sefirah* of kingdom. Similarly, among the 4 verses appearing in the triangle depicted to the right, the last one describes the last one describes **God's descent**, as it were, to earth for the first time since creation in order to give the Torah.

Let us analyze the mathematical properties of this structure. The bottom row of the triangle contains the entire fourth verse, the verse that precedes the Giving of the Torah.◀

Looking at the first and last words in the triangle, "In the beginning" (בְּרֵאשִׁית) and "saying" (לֵאמֹר), we find a clear allusion to the teaching of the sages that the Torah's first word—"In the beginning"—is also one of the Ten Utterances, or sayings of creation! The combined *gematria* of these two words is $1184 = 32 \cdot 37$.

The three corners (בְּרֵאשִׁית וַיְדַבֵּר לֵאמֹר) equal $1406 = 37$ times 38, the diamond (double triangle) form of 37! The two words at the beginning and end of the fourth, middle line of the triangle (הָאָרֶץ אִתּוֹ) equal 703, the triangle of 37 (and half of 1406)!

Constructing the 28 words of the four verses as a descending triangle from 7 to 1 (instead of from 1 to 7), the top line being the entire first verse of creation, gives the three corners (בְּרֵאשִׁית הָאָרֶץ לֵאמֹר) the value of $1480 = 40 \cdot 37$!

The midpoint of the first triangular form (from 1 to 7) is the word "from above" (מֵעַל) with reference to Abraham. The midpoint of the second triangular form (from 7 to 1) is the word "from above him" (מֵעָלָיו) with reference to Jacob, the word corresponding to "from above" in the previous verse. From these two words the sages learn that the Patriarchs are the Divine Chariot. Together they equal $296 = 8 \cdot 37$ (which equals the final word of the first verse, "the earth," הָאָרֶץ). $37 = yechidah$ (יְחִידָה), the single-one, the highest level of the soul that corresponds to the highest level of the Torah, the Working of the Divine Chariot, thus alluding to the fact that in essence all is one—"*Havayah* is One." We saw above that the value of all four verses is 6992, a multiple of 23, or *chayah* (חַיָּה), the living-one, thus alluding to the fact that all is included in the Working of Creation, and so both of the great mysteries of the Torah are in essence one. This is also indicated by the fact that the first verse of creation equals 2701, a multiple of 37, *yechidah*, yet the four words of the first verse that comprise the first and third lines of the above triangle from 1 to 7 (בְּרֵאשִׁית אֶת הַשָּׁמַיִם וְאֵת) = $2116 = 46^2$ (the average value of the four words is $529 = 23^2$!).

The Matriarchs

The Matriarchs, Sarah, Rebeccah, Leah, and Rachel, together with their husbands, were conduits for the essence of Divinity being revealed in the world. In fact, the sages state that Sarah was even greater than her husband Abraham in prophecy[23]; she was more in tune than her husband to God's will and purpose in creation and therefore knew that it was the right thing to banish Ishma'el and his mother from their household.

When we consider the Matriarchs together with the Patriarchs, our understanding shifts a bit. Earlier, we saw that the three general categories of Torah wisdom are: the Working of the Divine Chariot, the Working of Creation, and the debates of Abayei and Rava. These, we saw, are "the great thing," "the small thing," and the intermediate level between them. And so we may say that the Patriarchs correspond to the mystery of the Working of the Divine Chariot (men go to war, not women), the Matriarchs to the mystery of the Workings of Creation (the birth process of all reality from the Divine womb) and the 12 Tribes of Israel (the sons of Jacob) to

---

23. *Midrash Tanchuma, Shemot* 1.

the queries of Abayei and Rava (debates as to what exactly is the Divine will with regard to life's most minute details). Thus, the male seed (and the father principle in general) is likened to pure olive oil (the mystery of mysteries of the Torah, the Working of the Divine Chariot) and the female seed (and the mother principle in general) is likened to wine (the mystery of the Torah, the Working of Creation).[24]

7 Havayah's

We have just seen that 28 is a triangular number and that any 28 objects can be organized in the form of a triangle. Recall that at the start of this chapter we mentioned that a 1:4 word to letter ratio is considered the most ideal ratio in the Torah because *Havayah*, God's essential Name, has 4 letters. It follows that the most abstract and ideal 28 letters (which can subsequently be drawn in the form of a triangle) would be the 28 letters in 7 instances of *Havayah*, יהוה יהוה יהוה יהוה יהוה יהוה יהוה. The value of these is 182 = 7 · 26. Most importantly, 182 is the *gematria* of Jacob (יַעֲקֹב), the third (and choicest) of the Patriarchs. A verse in Isaiah alludes to this numerical relationship between *Havayah* and Jacob, "And now, so says *Havayah* who has created you Jacob…"[25] (וְעַתָּה כֹּה אָמַר יְהוה בֹּרַאֲךָ יַעֲקֹב). The sages[26] interpret that this verse should be read as God's calling to creation, "My world My world! Who has created you? Jacob has created you!" thus tying *Havayah* and Jacob together with creation.

Let us draw the 28 letters of the 7 *Havayah's* as a triangle,

```
                    י
                 ה    ו
              ה    י    ה
           ה    י    ו    ה
        ו    ה    י    ה    ו
     י    ה    ו    ה    י    ה
  ה    ו    ה    י    ה    ו    ה
```

24. See *Likutei Sichot*, v. 1, *Parashat Tetzaveh*.
25. Isaiah 43:1.
26. *Vayikra Rabbah* 36:4.

Let us highlight the letters in the triangle's three corners and midpoint.

<pre>
                          י
                     ה   ו
                ה   י   ה
           ו   ה   י   ה
      ה   י   ה   ו   ה   י
 ה   ו   ה   י   ה   ו   ה
</pre>

These four letters spell the word "shall be" (יִהְיֶה), alluding to the verse, "*Havayah* will be king over the whole earth; on that day *Havayah* shall be one and His Name will be one."[27] In Kabbalah, the word "shall be" (יִהְיֶה) is identified as the future spelling of God's essential Name, with the letter *vav* transformed into a *yud*, and alluding to the eternity of God's reign over the earth. The *yud* substituting the *vav* is the middle letter in the triangle, it is also the first letter of the 4th *Havayah* and thus corresponds with the fourth of the emotive *sefirot*, the *sefirah* of eternity (נֶצַח).

The 7 Names *Havayah* further allude to the first 7 explicit appearances of the Name *Havayah* in the Torah, all in the second account of creation. The seventh appearance is in the verse, "*Havayah Elokim* commanded man saying…" (...וַיְצַו יְ־הוה אֱ־לֹהִים עַל הָאָדָם לֵאמֹר).[28] This is the first explicit commandment in the Torah. The essence of the Torah is God's commandments as the word Torah itself means "instruction" and it conveys God's will in the form of His commandments to mankind. Indeed, the sages[29] find that this verse alludes to all 7 commandments that God commanded mankind, universally before the giving of the Torah.[30] Amazingly, the *gematria* of the verb "commanded" (וַיְצַו)—that appears here in juxtaposition to the full Name of *Havayah Elokim*—is itself 112, the same as the *gematria* of *Havayah Elokim* (יְ־הוה אֱ־לֹהִים)!

---

27. Zachariah 14:9.
28. Genesis 2:16.
29. *Sanhedrin* 56b.
30. About the 7 universal commandments, a.k.a., the Noahide Laws, see *Kabbalah and Meditation for the Nations*.

Because 112 = 4 · 28, the average value of the 4 letters of the verb "commanded" (וַיְצַו) is 28. This relates back to the four central and related verses discussed earlier, which share the same structure, 7 words and 28 letters, and which together have 112 letters.

Letter filling

One of the most important facets of *gematria* is letter filling. Each of the letters in the Hebrew alphabet has an exact name, and the letter's "filled" value (מִלּוּי) is thus equal to the value of its name. The first letter of a letter's name is always the letter itself. The other letters in the name are considered the letter's hidden or concealed dimension.

So let us see what this reveals for the word "commanded" (וַיְצַו). This word's four letters are *vav* (ו), *yud* (י), *tzadik* (צ), and *vav* (ו). Filling these letters (i.e., writing out there names) gives us, וו יוד צדיק וו, whose value is 248, the number of the positive commandments of the Torah and the value of "Abraham" (אַבְרָהָם).

The additional letters revealed through the filling process are ו וד דיק ו, revealing the concealed dimension of "commanded" (וַיְצַו) to equal 136. This number has a special affinity with the Name *Havayah*. It is the *gematria* of the word "voice" (קוֹל), the first word that appears in construct form with *Havayah* in the entire Bible, specifically in the phrase, "the voice of *Havayah*"[31] (קוֹל יְ־הוה). But, in this case the full name is used in the verse, and the complete construct form is "the voice of *Havayah Elokim*" (קוֹל יְ־הוה אֱ־לֹהִים), whose value too is 248!

What meaning can be gleaned from these two pieces of analysis? The first one reveals that before Adam's primordial sin, the Torah's 248 positive commandments were contained seminally in the revealed and concealed dimensions of God's single commandment to Adam, not to eat from the Tree of Knowledge of Good and Evil. After Adam sinned and transgressed this commandment, the 248

---

31. Genesis 3:8.

commandments were manifest once again in the "voice of *Havayah Elokim* moving through the garden…." [32]

An additioanl phase of our analysis is to look at the second filling of "commanded." To do this we expand each of the letters of the first filling, just as we did the first time. The result is, וו וו יוד ווד דלת צדיק דלת וו יוד דלת וו וו, which equals 1358. This is the exact value of the verse following the *Shema*, the essential statement of our monotheistic faith, which reads, "Blessed be the Name of the glory of His Kingdom forever and ever" (בָּרוּךְ שֵׁם כְּבוֹד מַלְכוּתוֹ לְעוֹלָם וָעֶד). Numerically, $1358 = 7 \cdot 194$, where 194 is the *gematria* of "justice," or "righteousness" (צֶדֶק). The same word is also the Hebrew name of the planet Jupiter, the planet associated in the Bible with Abraham. [33] All of God's commandments are directives for man to engage in deeds of righteousness (צֶדֶק) and charity (צְדָקָה, the feminine form of righteousness), on earth.

The *Havayah Elokim* verses

Let us further explore the first 7 instances of *Havayah* in the Torah by focusing on the verses in which they appear:

> *These are the chronicle of heaven and earth, having been created on the day that Havayah Elokim made earth and heaven.* [34]

> *No shrub of the field was yet on earth nor had any grass of the field yet sprouted because Havayah Elokim had not brought rain upon the earth, for there was no human to till the ground.* [35]

> *Havayah Elokim formed the human out of dust of the ground and blew into his nostrils a soul of life and the human became a living being.* [36]

32. For a great deal more on 248, see our website www.inner.org.
33. See Isaiah 41:2.
34. Genesis 2:4.
35. Ibid. v. 5.
36. Ibid. v. 7.

Havayah Elokim planted a garden in the eastern part of Eden and placed there the human whom He had formed.[37]

Havayah Elokim caused the ground to give forth every tree that is of pleasing appearance and good for consumption. The Tree of Life was in the center of the garden as was the Tree of Knowledge of good and evil.[38]

Havayah Elokim took the human and placed him in the Garden of Eden to cultivate and guard it.[39]

Havayah Elokim commanded the human saying, "Of every tree of the garden you may eat freely."[40]

The first thing to note is that in each verse, *Havayah* (which appears together with *Elokim*) is (in the Hebrew) preceded by a verb. Let us write out these phrases explicitly and the verb in the original Hebrew,

Havayah Elokim made	עָשׂוֹת
Havayah Elokim had [not] brought rain	הִמְטִיר
Havayah Elokim formed	וַיִּיצֶר
Havayah Elokim planted	וַיִּטַּע
Havayah Elokim caused the ground to give forth	וַיַּצְמַח
Havayah Elokim took	וַיִּקַּח
Havayah Elokim commanded	וַיְצַו

The values of all 7 verbs together is 1841. Adding the 7 Names *Havayah*, the 7 phrases◄ (*Havayah* and the associated verb) equal 2023,◄◄ the product of 7 and 289. Thus, the average value of each phrase is 289 = 17^2. 289 is also the *gematria* of the first mention of God's handiwork, "God created" (בָּרָא אֱ-לֹהִים). Thus, in a sense, each

37. Ibid. v. 8.
38. Ibid. v. 9.
39. Ibid. v. 15.
40. Ibid. v. 16.

► The *gematria* of the first two phrases is 1092 = 6 · 7 · 26, or 7 · 156, the value of "Joseph" (יוֹסֵף) or 6 · 182, the value of "Jacob" (יַעֲקֹב), or 42 · 26 (*Havayah*).

The *gematria* of the remaining five phrases is 931 = 7^2 · 19 (the value of the three dimensions of reality in *Sefer Yetzirah*, world, year, soul, עוֹלָם שָׁנָה נֶפֶשׁ).

►► One of the methods used in Kabbalah to understand words is to construct their **generative and degenerative forms**. As their names suggest, these two forms of the word follow the words coming into being and its disappearance from being. The generative form of the word "Torah" (תּוֹרָה) is

ת תו תור תורה

and the value of the entire generative form is 2023 (see text).

of the 7 actions performed by *Havayah* in the second account of creation is a detailed example of that which "God created" in the first account!

Here too we find that the 7 *Havayah's* exhibit a "golden division" (a topic we will explore more fully later) because in the Torah, the 7 *Havayah's* are separated by the description of the four rivers that emerged from the Garden of Eden. The first 5 phrases appear before this description and the final 2 follow it. ▸

The positional number of a word is defined as what number word it is from some point. From the beginning of the Torah, the positional numbers of the first 7 appearances of *Havayah* are: 477, 495, 514, 530, 542, 614, and 624. The only one of these numbers that is itself a multiple of 26 (i.e., the *gematria* of *Havayah*) is the last—"*Havayah* commanded"—$624 = 24 \cdot 26$.

But the sum of the positional numbers of all 7 appearances is,

$$477 + 495 + 514 + 530 + 542 + 614 + 624 = 3796$$

3796 is a multiple of 26, specifically, $3796 = 26 \cdot 146$, or *Havayah* times the *gematria* of "eternal" (עוֹלָם). The product of *Havayah* and "eternal" alludes to the verse referring to Abraham, "He proclaimed there in the Name of *Havayah* the eternal God"[41] (וַיִּקְרָא שָׁם בְּשֵׁם יְ־הוה אֵ־ל עוֹלָם). This phrase has a strong tie to the first verse of the Torah because its value is 1202, the same as the value of the Torah's first three words, "In the beginning God created" (בְּרֵאשִׁית בָּרָא אֱ־לֹהִים)! Furthermore, just God's description here as, "*Havayah*, the eternal God" (יְ־הוה אֵל עוֹלָם) is equal to 203, the *gematria* of "created" (בָּרָא). The value of "He proclaimed there in the Name of" (וַיִּקְרָא שָׁם בְּשֵׁם) is 999, also the *gematria* of "In the beginning… God" (בְּרֵאשִׁית אֱ־לֹהִים)! ▸▸

The end enwedged in the beginning

Let us complete this chapter by recalling that we started it by looking at the 1:4 words to letters ideal ratio. What we did not mention earlier

41. Genesis 21:33.

▸ In another interesting example of **inter-inclusion**, we find that of the 7 verbs, two are themselves multiples of 7:

the 5th, "made to grow" (וַיַּצְמַח) whose value is $154 = 7 \cdot 22$ and, the 7th, "commanded" (וַיְצַו), whose value is $112 = 7 \cdot 16$.

"Commanded" possesses another unique property. Its *gematria* (112) equals 7 times the value of its first two letters—*vav* and *yud* (וי)—which equal 16. Thus, "commanded" possesses an inner ratio of 1:6 between its first two letters (וי)—which grammatically form the prefix letters of the verb—and its final two letters (צו)—which grammatically are the verb's root letters.

▸▸ In the verse, "He proclaimed there in the Name of *Havayah* the eternal God," the word translated as "eternal" (עוֹלָם) also means "world" implying that God, *Havayah*, and the world are one and the same.

"World" (עוֹלָם) equals $146 = 2 \cdot 73$, where 73 is the value of "wisdom" (חָכְמָה), implying that the world is continuously being created with 2 aspects of God's wisdom, His **higher wisdom**, the Torah, and His **lower wisdom**, the wisdom of nature.

is that in the Torah's first verse the only word that exhibits the ideal 1:4 ratio (i.e., that has 4 letters) is "the earth" (הָאָרֶץ). Thus, the ratio of 1:4 (implicit throughout the first verse) resolves in a sense and becomes manifest at the end of the beginning of creation (indeed the first verse of creation serves as the heading or title of all of creation and includes all that is to come).

The first explicit appearance of *Havayah* in the Bible is in the first verse of the second account of creation, "These are the chronicles…"[42] Taking the Pentateuch as a whole, by far the most sweeping example of the end enwedged in the beginning would be to connect its final verse with its first verse. The concluding phrase of the Torah reads, "…before the eyes of all of Israel" (לְעֵינֵי כָּל יִשְׂרָאֵל) and its value is 761. When added to the *gematria* of the first verse of the Torah, "In the beginning God created the heavens and the earth" (בְּרֵאשִׁית בָּרָא אֱ־לֹהִים אֵת הַשָּׁמַיִם וְאֵת הָאָרֶץ), 2701, the sum is 3462, the exact numerical value of the first verse of the second account of creation, "These are the chronicles of heaven and earth, having been created on the day that *Havayah Elokim* made earth and heaven" (אֵלֶּה תוֹלְדוֹת הַשָּׁמַיִם וְהָאָרֶץ בְּהִבָּרְאָם בְּיוֹם עֲשׂוֹת יְהוָה אֱ־לֹהִים אֶרֶץ וְשָׁמָיִם), the first verse in which the Name *Havayah* appears.◄

► Sometimes this first verse of **the second account of creation** is actually considered to be the concluding verse of the first account, again following the principle of "the end is wedged into the beginning and the beginning is wedged into the end."◄

In the same vein, *Havayah* (י־הוה), the source for the ideal 1:4 ratio first appears in the Torah as the initial letters of the 4 consecutive words, "the sixth day. Heaven and earth were thus completed…"[43] (יוֹם הַשִּׁשִּׁי וַיְכֻלּוּ הַשָּׁמַיִם). These words conclude the Torah's description of the sixth day of creation and begin the description of the seventh day, the Shabbat, the conclusion and purpose of creation.

These are two examples where the end is enwedged in the beginning, i.e., that the end and the beginning are similar because they are, as it were, tied together like a circle.

As the all-inclusive verse of creation, the 7 words of "In the beginning God created the heavens and the earth" correspond

---

42. Genesis 2:4.
43. Genesis 1:31-2:1.

to the 7 days of creation themselves. Thus, the final, 7th word of the first verse, "the earth," corresponds to the 7th day of creation, the day of Shabbat, in whose account the Name *Havayah* first appears implicitly. The *gematria* of Shabbat (שַׁבָּת) is 702, or 27 times 26, where 26 is, once again, the value of *Havayah*. 702 is also the double-triangle, or diamond form of 26! The earth symbolizes creation reaching its lowest energy level, the secret of rest on the seventh day.

Havayah as a factor

At the outset of this chapter, we noted that God's essential four-letter Name, *Havayah*, does not appear in the first account of creation. Nonetheless, the value of *Havayah*, 26 is an important factor in it. Notably, beginning with the first word and skipping one word at a time, the *gematria* of the first, third, fifth, and seventh words of the verse (בְּרֵאשִׁית אֱ-לֹהִים הַשָּׁמַיִם הָאָרֶץ) is 1690, which is the product of 26 and 65.

65 itself is the *gematria* of God's holy Name pronounced *Adni* (אֲ-דֹנָי), the Name used to pronounce God's essential Name (as it is forbidden to pronounce the Teragrammaton as it is written, for which reason we refer to it as *Havayah*, using a permutation of its letters).

Both 26 and 65 are multiples of 13. $26 = 2 \cdot 13$ and $65 = 5 \cdot 13$, so their combined *gematria* is $7 \cdot 13 = 91$, the triangle of 13. The sectioning of 7 into 2 and 5 is called the "golden division" (related to, but not exactly the same as the Golden Section). This is because the letters of the word "gold" (זָהָב) in Hebrew represent the equation: $7 = 5 + 2$.

We have found that 13 is the common factor of the two inter-related Names of God, *Havayah* and *Adni*. 13 is the value of "one" (אֶחָד) and "love" (אַהֲבָה). Since 1690 is $10 \cdot 169$, it follows that not only is 13 a factor of 1690, but that $13^2 = 169$ is also! Furthermore, the *gematria* of the initial letters of these four words (whose sum is 1690), ב א ה ה, is

▶ Note that the *gematria* of the first and last words of the verse (בְּרֵאשִׁית הָאָרֶץ) is 1209 = 13 · 93, while the *gematria* of the third and the fifth words of the verse (אֱ-לֹהִים הַשָּׁמַיִם) is 481 = 13 · 37.

itself 13. In fact, these initials permute to spell the word "love" (אַהֲבָה). ◄

In passing, let us note that apart from the initials, the sum of the remaining letters (ראשית להים שמים ארץ) is 1677, which is one and one-half times times 1118—the value of the *Shema*, "Hear O' Israel *Havayah* is our God, *Havayah* is one" (שְׁמַע יִשְׂרָאֵל יְ-הוָה אֱ-לֹהֵינוּ יְ-הוָה אֶחָד). As explained elsewhere in length, 1118 is the smallest number that is a multiple of both 26 (*Havayah*) and 86, the value of *Elokim* (אֱ-לֹהִים), the Name of God appearing in the first verse of Genesis.

THE PRIMORDIAL TRIANGLE

"All sevens are endeared"

The first verse of the Torah contains 7 words and 28 letters. The sages state that "all sevens are endeared"[1]; Shabbat is the seventh day of the week, Chanoch (Enoch) is the seventh generation from Adam, Moses is the seventh generation from Abraham, etc.

We have made use of the figure of the triangle in chapter 1. Now let us define it more rigorously. The triangular value of any positive integer number n is defined as the sum of integers from 1 to that number n. Or stated in mathematical form,

$$\triangle n = \sum_{k=1}^{n} k$$

It is easy to see that since 28 is the sum of integers from 1 to 7 (1 + 2 + 3 + 4 + 5 + 6 + 7) that 28 = △7. 28 happens to also have another special relationship with the number 7, since it is a multiple of 7. 28 = 4 · 7. So 28 is both a multiple of 7 and the triangle of 7. This is not a coincidence, because 4 is actually *the midpoint* of 7. To understand this concept we need to backtrack a little.

If we wish to calculate the triangle of a number n quickly, without having to add all the integers up to that number, we can use a simple formula (which is actually quite easy to come by if you think in geometric terms, but we will not be getting into that here):

$$\triangle n = \frac{n\,(n+1)}{2}$$

---

1. *Vayikra Rabah* 29:11.

So in the case of the triangle of 7, we get:

$$\triangle 7 = \frac{7\,(7 \pm 1)}{2} = \frac{7 \cdot 8}{2} = 7 \cdot 4$$

So using the formula for quickly calculating the triangle of 7, we find that it is simply the product of 7 and 4, which brings us back to 4 being the midpoint of 7. If you are a little mathematically inclined you can see that this formula is telling us that if n is an odd number, you find its triangular value by multiplying n by its midpoint.

All odd numbers have a midpoint. Think of 7 coins placed in a row. Which coin is in the middle? The fourth! (It should be clear that even numbers do not have a midpoint◄). So 4 is the midpoint of 7. To state this mathematically,

> When m *is an odd number, we use the notation* n <m *to denote that "n is the midpoint of m," where by definition,*
>
> $$n = \triangleleft\,(2n - 1)$$

When we set n = 4, we get that 4 <7. Now, how about finding the midpoint of an odd number directly? All we have to do is work in the other direction and we can write that if n is an odd number then its midpoint is found by adding 1 to n and dividing by 2.

> The midpoint of a positive odd integer can be calculated using the formula:
>
> $$\triangleleft n = \frac{n \pm 1}{2}$$

But, now note that this is part of the right-side expression we used to calculate the triangle of an integer number n, which through substitution now becomes,

> The triangular value of a positive odd integer is given by the expression,
>
> $$\triangle n = n(\triangleleft n)$$

► Mathematically, the midpoint is to **odd numbers** as division by 2 is to even numbers.

Returning to the case of n = 7, since 7 is odd and its midpoint is 4, it follows that $\triangle 7 = 7 \cdot 4$.

On our way to seeing that the first verse of the Torah can be drawn as a triangle, we have learnt some basic principles of triangular numbers.

So, now let us figure the first verse, "In the beginning, God created the heavens and the earth" (בְּרֵאשִׁית בָּרָא אֱ-לֹהִים אֵת הַשָּׁמַיִם וְאֵת הָאָרֶץ) into a triangle of 7. We get the following,

```
              ב
           ר   א
        שׁ   י   ת
     א   א   ר   ב
  א   ם   י   ה   ל
ם   י   מ   שׁ   ה   ת
ץ   ר   א   ה   ת   ו
```

First we observe that the 6 letters of the first word (בְּרֵאשִׁית) make up the first three lines of the triangle. This is of course because $6 = \triangle 3$ and gives the following triangle,►

```
        ב
     ר   א
  ת   י   שׁ
```

Now note that the bottom line of the triangle—its base level (or "earth")—is populated by the letters of the final phrase, "…and the earth" (וְאֵת הָאָרֶץ).

Actually, the first verse is so triangular in its essence—a property indicating that it is "evolutionary"—that out of its words, taken in sequential order, we can perfectly form the first 7 triangles: 1, 3, 6, 10, 15, 21, and 28. We have already seen 6 and 28. Let us now illustrate the other 5.

1) Out of the word "He created" (בָּרָא) we can form $\triangle 2 = 3$,

```
     ב
  ר   א
```

► Analyzing the triangle, we note that the six letters of **the Torah's first word**, "In the beginning" (בְּרֵאשִׁית), are divided into two groups of 3. The first 3 letters (ברא) are in the triangle's top two rows and the final 3 in its base (שית). The division of the Torah's first word of the Torah into two words, each with 3 letters is found in the *Zohar* (I, 3b), which explains that the two halves, ברא and שית, mean, "He created six" (בָּרָא שִׁית) alluding to the six days of creation (on the seventh day He rested).

The number 6 in this context also refers to the 6 permutations of the 3 letters of the word "He created" (בָּרָא) itself (it follows then that each permutation of "He created" corresponds to one of the days of creation).

We will explore this topic in greater depth in chapter 9.

2) The words "the heavens and" (אֶת הַשָּׁמַיִם וְאֶת) have 10 letters, which we form into △4 = 10,

```
            א
          ת   ה
        ש   מ   י
      ת   א   ו   ם
```

3) The words "God created the heavens" (בָּרָא אֱ־לֹהִים אֶת הַשָּׁמַיִם) have 15 letters, which we form into △5 = 15,

```
            ב
          ר   א
        א   ל   ה
      ת   א   ם   י
    ה   ש   מ   י   ם
```

4) The words, "In the beginning, God created the heavens," (בְּרֵאשִׁית בָּרָא אֱ־לֹהִים אֶת הַשָּׁמַיִם) contain 21 letters, allowing us to shape them into △6 = 21,

```
            ב
          ר   א
        ת   י   ש
      ב   ר   א   א
    ל   ה   ם   י   א
  ת   ה   ש   מ   י   ם
```

5) Finally, in the Torah scroll, the first letter of the first word, "In the beginning" (בְּרֵאשִׁית) is a large letter *beit* (ב), written larger than all the other letters. In the entire Bible, each of the letters of the *alef-beit* appears once as a big letter and once as a small letter. Additionally, the *beit* of "In the beginning" is a prefix letter, not one of the word's root letters (which are ראש, meaning "head"). Thus, the first letter of the Torah actually stands apart, transcending the remaining letters of the first word and the first verse. Since it is a single letter, it can be formed into △1 = 1,

```
            ב
```

If we take a look back, we will see that since the first 3 letters of the word, "In the beginning" (בְּרֵאשִׁית) are the same as the second word, "created" (בָּרָא), then from the first word alone we can form the triangles of 1, 2, and 3. Another way of seeing this is that looking at the first three rows, they can be broken down, meaningfully—as just explained—into the triangles of 1, 2, and 3, ▶

```
  ב          ב          ב
           ר א        ר א
                     ש י ת
```

Now, the sum of the first three triangular numbers, 1, 3, and 6 is 10, which is itself the triangle of 4,

$$\triangle1 + \triangle2 + \triangle3 = \triangle4$$

10 is also a tetrahedral number, whose rigorous treatment we leave for another opportunity. Let us just define mathematically that,

> The tetrahedron of a positive integer n
> is denoted by the symbol ⧍n and is defined as,
>
> $$⧍n = \sum_{k=0}^{n} \triangle k$$

Thus, the sum of the first 3 triangular numbers, contained with the first word of the Torah as above, is the tetrahedron of 3! It therefore follows that,

$$\triangle4 = ⧍3$$

This particular finding, that 10 is both the triangle of 4 and the tetrahedron of 3 is one of the many reasons that numbers in the Torah are written using base 10, making the decimal system the Divine choice for expressing quantities.

Now, going back to the fact that the first verse is formed into a

▶ Incredibly, the value of the letters in these 3 triangles, בראשית ברא ב, is **1118**, the value of the *Shema*, the verse (Deuteronomy 6:4) that forms the essential declaration of faith in God's oneness (שְׁמַע יִשְׂרָאֵל יְיָהוֹה אֱיָ לֵהֵינוּ יְיָהוֹה אֶחָד).

Building upon this idea, the next three triangles formed from the first verse's letter will be,

```
                                          ב
                           ב            ר א
            ב            ר א          ש י ת
          ר א          ש י ת        ב ר א א
        ש י ת        ב ר א א      ל ה י מ א
      ב ר א א      ל ה י מ א    ת ה ש מ י ם
```

The value of the letters in these 3 triangles is 5436, or 2 times 2718, where 2718 is the value of the 3 verses that make up the Priestly Blessing (Numbers 6:24-26), "May God bless you and keep you. May God shine His countenance upon you and give you grace. May God lift His face towards you and grant you peace" (יְבָרֶכְךָ יְיָהוֹה וְיִשְׁמְרֶךָ. יָאֵר יְיָהוֹה פָּנָיו אֵלֶיךָ וִיחֻנֶּךָ. יִשָּׂא יְיָהוֹה פָּנָיו אֵלֶיךָ וְיָשֵׂם לְךָ שָׁלוֹם).

triangle with 7 rows, let us explain that these 7 rows correspond to the seven emotive *sefirot*—loving-kindness, might, beauty, victory, acknowledgment, foundation, and kingdom. The 7 rows also correspond to the seven days of creation.

It follows then that the first three lines correspond to the three primary emotive *sefirot*—loving-kindness, might, and beauty—whose archetypal souls are the three Patriarchs, Abraham, Issac, and Jacob. There is a beautiful numerical point at the end of all this (which actually serves as the starting point for a whole new meditation on the verse, one that we leave for another time). The first three rows, as we have seen, are fully populated by the verse's first word, "In the beginning" (בְּרֵאשִׁית). Its numerical value is 913◄, which is also the value of the phrase, "three things together" (שְׁלֹשָׁה דְּבָרִים יַחַד),[2] referring to the three Patriarchs alluded to by the first three rows of our triangle.

Let us take another moment to dwell on this beautiful phrase. It contains three words. The first word, "three" (שְׁלֹשָׁה) equals 635. The first two words taken together form the phrase "three things"◄ (שְׁלֹשָׁה דְּבָרִים) whose value is 891. All three words, taken together, "three things together" (שְׁלֹשָׁה דְּבָרִים יַחַד) have the value 913. The average value of these three sets of words is then: $635 + 891 + 913 = 2439$, and $2439/3 = 813$◄◄, the value of the Torah's third (!) verse, "God said let there be light and there was light" (וַיֹּאמֶר אֱ־לֹהִים יְהִי אוֹר וַיְהִי אוֹר).

A very triangular verse

The *gematria*, i.e., the numerical value of the entire verse, "In the beginning, God created the heavens and the earth" (בְּרֵאשִׁית בָּרָא אֱ־לֹהִים אֵת הַשָּׁמַיִם וְאֵת הָאָרֶץ) is 2701, which itself is a triangular number. 2701 is the triangle of 73 ($2701 = \triangle 73$), and since 73 is an odd number this means that it is the product of 73 and its midpoint, 37 ($2701 = 37 \cdot 73$). 73 is the *gematria* of "wisdom" (חָכְמָה). God created

▶ These two words, "three things" (שְׁלֹשָׁה דְּבָרִים), can also be read as "three leaders" (שְׁלֹשָׁה דַּבָּרִים).

Amazingly, their numerical value (which of course does not change with the change in vowelization), 891, is also the *gematria* of the names of the three leaders who took us out of Egypt, "**Moses, Aaron, Miriam**" (מֹשֶׁה אַהֲרֹן מִרְיָם). Indeed, Moses, Aaron, and Miriam correspond to the three Patriarchs: Moses to Jacob (following the *Zohar*'s statement, that "Jacob externally, Moses internally"), Aaron, the man of loving-kindness, to Abraham the archetypal soul of loving-kindness, and Miriam to Isaac who is considered a soul from the feminine reality.

We will delve into this phrase and its various meanings in greater depth in chapter 7.

▶▶ **813** is also the value of key phrases in the Torah's account of creation, "God separated the light from the darkness…" (וַיַּבְדֵּל אֱ־לֹהִים בֵּין הָאוֹר וּבֵין הַחֹשֶׁךְ) and the phrase, "God said, 'Let us make a man'" (וַיֹּאמֶר אֱ־לֹהִים נַעֲשֶׂה אָדָם).

---

2. From Rabbi Avraham Abulafia.

the world with wisdom.▶ 37 is also the *gematria* of "wisdom" in ordinal numbering.

> *Ordinal numbering (מִסְפָּר סְדּוּרִי) assigns each letter with the value of its ordinal position in the Hebrew alphabet.*

In the case of "wisdom" (חָכְמָה),

- ♦ ח = 8
- ♦ כ = 11
- ♦ מ = 13
- ♦ ה = 5

And their sum is of course 37.

Now, let us calculate the *gematria* of the base of the triangular form of the first verse alone. The letters there are,

$$וְאֵת הָאָרֶץ = 703$$

Since in Hebrew the name of the first letter, *alef* (א), whose value is 1, is the same as the word "one thousand" (אֶלֶף), it is common to treat 1000 as being equal to 1.▶▶

In our case the 2 thousands of 2701 (the value of the entire verse) would then reduce to 2 and when added to the remainder, 701, would equal 703, the value of these final two words, "and the earth," (וְאֵת הָאָרֶץ) which form the base row of our triangle! This equivalency suggests that the entire triangle projects itself onto its bottom row, the 7th row, which corresponds to the *sefirah* of kingdom, the earth, the spiritual attribute of lowliness.

Before continuing this train of thought, Let us note that 703 too is a triangular number.

$$703 = \triangle 37$$

But, so is the value of "kingdom" (מַלְכוּת),

$$496 = \triangle 31$$

The difference between 703 (the words in the "kingdom" row of our

▶ There are three separate sacred translations of the Torah into Aramaic known as: *Onkelos, Yerushalmi*, and *Yonatan [ben Uzi'el]*.

The *Yerushalmi* translates the first word of the Torah, "**In the beginning**" (בְּרֵאשִׁית) as, "With wisdom" (בְּחוּכְמָא). According to this translation, the first verse reads, "With wisdom God created the heavens and the earth."

In Hebrew, the first letter of the verse, the *beit* (ב) means either "in" or "with." Thus, "In the beginning..." can also read as, "With the beginning...." The word "beginning" (רֵאשִׁית) alludes to the phrase "The beginning of wisdom" (רֵאשִׁית חָכְמָה). This phrase appears twice in the Bible, once in the verse (Psalms 111:10), "The beginning of wisdom is the fear of God..." (רֵאשִׁית חָכְמָה יִרְאַת הוי׳) and once in the verse (Proverbs 4:7), "The beginning of wisdom—acquire wisdom..." (רֵאשִׁית חָכְמָה קְנֵה חָכְמָה).

Since there are two instances of this phrase, "the beginning of wisdom," in the Bible, the value of the letter *beit*, 2, can be understood as referring to them.

▶▶ The verse that alludes to this phenomenon—of 1000 being treated as 1—is (Deuteronomy 32:30), "**How does one pursue a thousand**" (אֵיכָה יִרְדֹּף אֶחָד אֶלֶף).

▶ While 703 is the triangle of 37, the difference between 703 and 496—the value of "kingdom" (מלכות)—is 207, a multiple of 23, illustrating once again how **37 and 23** are companion numbers.

▶▶ There is another form of numbering known as "final reduced numbering" (מספר קטן אחרון). The value of a word in this system is simply the sum of the digits of the reduced value, taken recursively until the sum is between 1 and 9. In the case of "wisdom," since the reduced value is 19, adding the digits gives 10, which when added once more, yield 1.

Taking all **four levels of wisdom**, 73, 37, 19, and 1, their sum is 130, the value of the word "eye" (עין). Of all the organs in the body, the eyes are the ones most associated with wisdom, as in "the eyes of the congregation" (Numbers 15:24) a connotation for the wise elders and, "the wise see [what will come to pass]" (Tamid 32a).

If we add to this 10, the final midpoint and the reduced value, the sum becomes 140, which is the value of "wisdom" and "understanding" (חכמה בינה), "sun" and "moon" (חמה לבנה), and more.

triangle) and 496 (the value of "kingdom") is 207 = 9 · 23.◄ 207 is the *gematria* of some very related terms in Kabbalah, among them, "light" (אור) and "infinity," or "the Infinite One" (אין סוף). Thus the concluding phrase of the first verse of the Torah, the bottom, kingdom-row of our triangle equals the value of the phrase, "light of kingdom" (אור מלכות) or alternately, "kingdom of the Infinite One" (מלכות אין סוף).

Following our earlier definition regarding triangles and midpoints,

$$703 = \triangle 37 = 37 \cdot 19$$

where 19 is, by definition the midpoint of 37.

19 itself is the *gematria* of "wisdom" (חָכְמָה) in reduced numbering.

> In reduced numbering (מִסְפָּר קָטָן) the values of the letters from alef to tet retain their normative value. The letters from yud and on, the value becomes the normative value less the trailing zeros (i.e., mod 9).

Thus, the reduced values◄◄ for the four letters of "wisdom" (חָכְמָה) are,

letter	normative value	reduced value
ח	8	8
כ	20	2
מ	40	4
ה	5	5
sum in reduced value		19

To summarize what we have found,

$$19 < 37 < 73$$

Or, in words, "wisdom" (in reduced value) is the midpoint of "wisdom" in ordinal value, which in turn is the midpoint of "wisdom" in normative value.

Midpoint series

As stated earlier, every odd number has a midpoint. If that midpoint is itself an odd number, it too has a midpoint, and so on. The process continues until reaching a midpoint which is an even number, because even numbers do not have a midpoint. In our case, if we continue we find that ▶,

$$10 < 19 < 37 < 73 ▶▶$$

This is of course a unique series (and none of its numbers will appear in any other series of midpoints). In fact, it is easy to convince ourselves that every even number produces such a series. ▶▶▶ The even number beginning the series of midpoints is called sometimes called the origin of the midpoint series (in a moment we will look at it from a different perspective). In our case, 10 is the value of the letter *yud* (י), which we recall is the middle element in the triangular form of the Torah's first verse; we refer to it as the "central geometric element," and will have more to say about it later.

One of the mathematical properties of triangular numbers is that every third one (the triangles of 1, 4, 7, 10, 13…) has a central geometric element. What this means is that the triangular shape of every third triangle can be divided into concentric triangles situated around a geometric center.

Let us state this rule in mathematical terms,

> *The triangle of every integer of the form 3n ± 1*
> *has a central geometric element.*
> *(Note that the expression 3n ± 1 is equivalent*
> *to the residue class of 1 mod 3.)*

In our particular case, because 7 is expressible as 3n ± 1 (where n = 2), the triangle of 7 has a central geometric point. Let us depict this graphically by coloring the concentric triangles around the central geometric element in different shades of grey:

▶ Note that 10, the final midpoint of the "wisdom" series is the sum of digits of 73, 37, and 19.

▶▶ For mathematical reasons, we usually extend series we are meditating on to **13 places**. Doing so for the midpoint series beginning with 10, we find that the full series to 13 places is: 10, 19, 37, 73, 145, 289, 577, 1153, 2305, 4609, 9217, 18433, 36865. We expect to be able to find many points of interest relating these numbers to the topic at hand. For instance, 289 is the value of the second and third words in the Torah's first verse, "…God created" (בָּרָא אֱלֹהִים). The next number 577 is the value of the three letters of the word "one" (אֶחָד) each raised to the third power: $1^3 + 8^3 + 4^3 = 577$. Most beautifully, the 13th number, 36865 is a multiple of 73, the value of "wisdom" (חָכְמָה).

▶▶▶ Midpoint series present a beautiful illustration of **the paradoxical nature of infinity**, because every even number produces a unique series of odd numbers, going on for infinity.

Incidentally, we can show that there is nothing mysterious about the even numbers in this respect by simply either adding 1 or subtracting 1 from all the numbers in the midpoint series produced by even numbers. What we will get is that every odd number produces, in a sense, a unique infinite series of even numbers.

```
                    ב
                  ר  א
                ש  י  ת
              ב  ר  א
            א  ל  ה  י  ם
          ם  י  מ  ש  ה  ת
        ו  א  ת  ה  ר  א  ץ
```

We find that the letter *yud* (י) of *Elokim*◄ is the geometrically centered element of the triangular form of "In the beginning, God created the heavens and the earth."

In Kabbalah, the letter *yud* (in particular the *yud* of God's essential Name, *Havayah*, its first letter, which unites with the *yud* of *Elokim*—the Name of God that appears in the first verse) itself corresponds to the *sefirah* of "wisdom."

Let us say a bit more about the special relationship between *Elokim* and *Havayah*, focusing on the *yud* of *Elokim*, which alludes to the *yud* of the Name *Havayah*. As noted already in the previous chapter, in the first account of creation only the Name *Elokim* appears. The *gematria* of *Elokim* (אֱ־לֹהִים) is 86 which equals "nature" (הַטֶּבַע), indicating that the Name *Elokim* refers to God's light and power as expressed in nature, as one with nature. *Havayah* refers to the supernatural, infinite light and power of the Creator. *Havayah* is the Name that works miracles that defy the laws and norms of the natural order. The hidden presence of *Havayah* within *Elokim*—as alluded to by the letter *yud* in *Elokim*—means that nature itself is in essence supernatural, that nature is ultimately the miracle of miracles.◄◄

There are two letters common to both *Havayah* and *Elokim*: the *yud* (י) and the *hei* (ה). The common *yud*, whose shape is referred to as a "formed point," a point of contraction and concentrated energy, unites the two Names, while the common *hei* reveals the individuality of each Name.◄◄◄

▶ To see that the *yud* of *Elokim* is indeed **the middle letter** of the triangle count from each of the three corners of the triangle until you reach the letter *yud*. You will find that it is the 13th letter from each corner.

13 is the *gematria* of "one" (אֶחָד), "*Havayah* is one." The middle points of those triangles that possess middle points (every third triangle, as noted above) comprise the quadratic series of inspirational, or interface numbers: 1, 5, 13, 25, 41, 61, 85, ….

▶▶ On a deeper level, the hidden presence of *Havayah* within *Elokim* is the secret of the **contraction and concentration** (called the *tzimtzum*, a word that carries both meanings) of the infinite (*Havayah*) within the finite (*Elokim*, which equals "nature") and of the absolute unity ("*Havayah* is one"—the culmination of the *Shema*) within the origin of plurality (grammatically, *Elokim* is in the plural form).

▶▶▶ The four letters of God's essential Name, *Havayah*, allude to **the process of creation**. The *yud* alludes to contraction and concentration, the first *hei* to an initial expansion (formation of plurality), the *vav* to the drawing down of light, and the final *hei* to yet a second expansion. As such, the *yud* that is common to both *Havayah* and *Elokim* symbolizes a state of concentrated unity, a state in which both Names are still unified, while their common *hei* symbolizes a state of relative plurality, where each Name takes on a separate meaning, *Havayah* representing the super-natural and *Elokim* representing the natural.

The triangle's heart

The central geometric element of a triangle can be considered its seminal point, the heart that generates the triangle from within. In our case, this is the letter *yud* (י) = 10, which is a triangular number itself,

$$10 = י = \triangle 4 \blacktriangleright$$

Around the seminal *yud* of the triangle there are 27 (= 3^3) letters:

```
            ב
          ר  א
        ש  י  ת
      א  ר  ב
    א  מ      ה  ל
  ם  י  ה  ש  ת
ו  א  ה  ת  ר  ץ
```

Their *gematria* is 2691 = 13 · 207. 207 is the value of "light," (אוֹר), the object of creation on the first day. Thus, the triangle of the first verse suggests that the seminal *yud* creates around it 13 "lights." Indeed, in Kabbalah we are taught that 13 lights emanate from the supernal *sefirah* of crown (the secret of the Torah's first verse). Here we see that they emanate from the seminal *yud* of *Elokim*, the central geometric element of the triangle of the first verse.

Note that "light" (אוֹר) itself, 207, is equal to 9 · 23, where 23 is the value of "radiance" (זִיו)▸▸ ; the full spectrum of light is thus understood to possess 9 measures, or lines of radiance. Throughout the Torah, 23 is the companion number to 37.▼▼▼

▸ This relationship is alluded to in **the letter *yud*'s filling**, which is יוד. The *gematrias* of the two letters that fill the *yud* (*vav* and *dalet*) are 6 and 4 respectively. 6 ⊥ 4 = 10, meaning that the filling of the *yud* equals itself. 6 is the triangle of 3, and then comes 4, the bottom line completing the triangle of 4.

▸▸ In Hebrew, there are 13 synonyms for "light." They correspond to the 13 Measures of Divine Mercy (מִדּוֹת הָרַחֲמִים), that emanate in the form of the 13 *Tikunei Dikna* (13 enclothements of the beard) from the *sefirah* of crown, in the following manner:

tikun	synonym	tikun	synonym
אֵ-ל	אוֹר	נֹצֵר חֶסֶד	חַשְׁמַל
רַחוּם	בֹּהַר	לָאֲלָפִים	יָקָר
וְחַנּוּן	זֹהַר	נֹשֵׂא עָוֹן	בֹּהַק
אֶרֶךְ	נֹהַר	וָפֶשַׁע	זֶרַח
אַפַּיִם	נֹגַהּ	וְחַטָּאָה	הִלָּל
וְרַב חֶסֶד	צֹהַר	וְנַקֵּה	טֹהַר
וֶאֱמֶת	זִיו		

According to this order, the word אור (whose value is 207) is the first and the word זיו (whose value is 23) is the seventh.

▸▸▸ The companion numbers 23 and 37 are also the **golden section** of 60, which is defined as the division of an integer n into two integers whose sum is n and whose ratio (when the larger is divided by the smaller) is as close as possible to 1.618

In practice, to find the golden section of a number n, divide it by 1.618 and round to the closest integer. The second integer is found by subtracting the rounded result from the original number n.

Another example of how 23 and 37 appear together is in regard to the two emotions that according to the *Tanya* exist simultaneously in the Jewish heart, "joy" (חֶדְוָה) and "crying" (בְּכִיָה) (*Tanya*, ch. 34, based on *Zohar* II, 255a). The value of the first is 23, that of the second, 37.

Regarding the first verse of the Torah, its total *gematria* is 2701, a multiple of 37, while the value of the first, middle, and final words, "In the beginning... [the]... the earth" (בְּרֵאשִׁית...אֶת...הָאָרֶץ) is 1610, a multiple of 23.

The first concentric triangle surrounding the seminal *yud* of the complete triangle comprises 9 letters. The second concentric triangle comprises 18 letters. Thus, the second triangle (the external frame of the entire figure) surrounds 10 (the value of *yud*) letters.

These 10 letters are,

$$\begin{array}{ccccc} & & \text{י} & & \\ & \text{ר} & & \text{א} & \\ \text{ה} & & \text{י} & & \text{ם} \\ \text{ה} & \text{ש} & & \text{מ} & \text{י} \end{array}$$

> ▶ **621** is the value of "crown" (כִּתְרָא) in Aramaic, which itself possesses 3 super-conscious lights (from which emanate 13 lights). In the soul, the three lights of the super-conscious are faith (אֱמוּנָה), pleasure (תַּעֲנוּג), and will (רָצוֹן).

and, their *gematria* is $621 = 3 \cdot 207$, "light" (אוֹר), as above. ◀

The 9 letters in the first concentric triangle around the seminal *yud* are,

$$\begin{array}{ccccc} & & \text{י} & & \\ & \text{ר} & & \text{א} & \\ & \text{ם} & & \text{ה} & \\ \text{ה} & \text{ש} & & \text{מ} & \text{י} \end{array}$$

and their *gematria* is 611, also the *gematria* of "Torah" (תּוֹרָה)—"the Torah is light."[3] Also, $611 = 13 \cdot 47$.

A word about this product: 13 is the value of "love" (אַהֲבָה) and 47 the value of "selflessness" (בִּטוּל). These are the experiential aspects of the *sefirot* of loving-kindness and wisdom, respectively. Thus, the Torah particularly acts to unify two attributes of the soul. Loving-kindness lies exactly under wisdom on the right axis of the Tree of Life—the planar model of the *sefirot*.[4] Since Abraham, the first Jew, is the archetypal soul of loving-kindness, the Divine light that emanates from wisdom into Abraham's soul root is referred to in Kabbalah as, "the God of Abraham."[5]

Both 13 and 47 are prime numbers. 13 is the 7th prime, 47 is the 16th. ◀◀

> ▶▶ For centuries, mathematicians considered the number "one" the first of the **prime numbers**. In the past few decades, for various technical reasons, "one" has been estranged from the prime numbers and placed in a class of its own. However, in Kabbalah, it is still considered the first prime.
>
> This is not a matter of dispute, but rather of convention. Both viewpoints are valid.

---

3. Proverbs 23:6.
4. See in length, *What You Need to Know About Kabbalah*, p. 82.
5. Genesis 26:24.

Let us define the "source" (מָקוֹר) of a number, as follows:

> The "source" of a number n is the product
> of the ordinal values of its prime factors.

In our case, the source of Torah (611, תּוֹרָה), is $7 \cdot 16 = 112$, which we have already seen is the full Name of God appearing in the second account of creation, *Havayah Elokim*.[6]

The second concentric triangle is,

```
                ב
           ר         א
        ת                ש
     א                      ב
   א                          ל
  ם                            ת
 ץ    ר    א    ה    ת    א    ו
```

The *gematria* of its 18 letters is $2080 = 80 \cdot 26$ (*Havayah*) $= \triangle 64$. 2080 is also $10 \cdot 208$, where 208 is the value of "Isaac" (יִצְחָק). 2080 thus symbolizes the revelation of Isaac's soul-root through all ten powers of the soul. But, 64 also relates to Isaac, since Isaac is considered the chariot for God's attribute of "judgment" (דִין) whose *gematria* is 64. Judgment is particularly related to the Torah's first verse, since the sages relate that initially, God's intent was to create a world controlled by His attribute of judgment.[7]

The first three rows of the second concentric triangle are also a triangular number. They are,

```
        ב
    ר       א
  ת           ש
```

And the values of these letters is $903 = \triangle 42$

---

6. See above, in chapter 1.
7. See *Rashi* on Genesis 1:1.

Note that in this figure, the value of the letters in the fourth line

א   ב

is 3 and $3 = \triangle 2$.

Also, the *gematria* of the letters in the third and fourth lines together,

ת   ש

א   ב

is 703, exactly the value of the bottom line (וְאֵת הָאָרֶץ), 703, which as we saw above is equal to $\triangle 37$!◄

▶ Considering that the value of the entire first verse is 2701 or $\triangle 73$, we can make another interesting observation about **the central geometric element**. 73 is also expressible as 3n ± 1 (in this case, n = 24), which means that the triangle of 73 has a central geometric element. In this case, it would be 1201, which is also the 25th inspirational number (a topic we will cover some more in the next chapter).

Now, the first half (as defined by the Torah's cantillation marks) of the Torah's first verse is, "In the beginning God created" (בְּרֵאשִׁית בָּרָא אֱ־לֹהִים), whose value is 1202, one more than the value of the middle point of the value of the triangle formed from the entire verse.

MIDPOINTS AND CREATION

Nothingness, truth, and the midpoint

Rabbi Isaac Luria, the Arizal, began his description of creation with the statement that the world was created from "the middle point" (נְקוּדָה אֶמְצָעִית), as it were, of God's infinite light.[1] The middle point is the point of "nothingness" (אַיִן), from which the world is continuously being created *ex nihilo*, and the point of "truth" (אֱמֶת), from which the Torah, God's will, is made known to man. Every individual world or consciousness within creation has its own middle point, its point of nothing and truth. All the middle points of all the worlds connect to one another, ultimately connecting to the primordial middle point, the point where God first "contracted" (צִמְצֵם) His infinite light. This creates what can be likened to the modern physical concept of a black hole singularity, a wormhole that connects pre-creation with creation.[2]

Let us look at these three related concepts—nothingness, middle, and truth—from a mathematical perspective. The *gematria* of nothingness (אַיִן) is 61, of "middle" (אֶמְצַע), 201, and of "truth" (אֱמֶת), 441. Their sum is 703, which as we have seen in the previous chapter is the value of the bottom row of the triangle formed from the Torah's first verse and containing the words, "and the earth" (וְאֵת הָאָרֶץ). We also saw that $703 = \triangle 37$. Incredibly, when we write these 3 words—nothingness, middle, and truth—using letter filling, yielding, אלף יוד נון אלף מם צדיק עין אלף מם תיו, their value is $1369 = 37^2$!▶

Whenever we have exactly 3 integer values, we can use them to develop a quadratic series (we leave the derivation for another time). To treat a set of 3 integers as a series and to extend the series

---

1. *Etz Chaim* 1:2.
2. *Tanya*, end of ch. 13. The Alter Rebbe mentions all three words in this passage.

▶ One of the basic algebraic expressions coming out of figurate numbers is that **the sum of two consecutive triangles** is a square. Or, in mathematical notation,

$$\triangle n + \triangle(n - 1) = n^2$$

This can be illustrated geometrically by dividing a square number into two triangles. Take for instance 4^2 dots ordered in a square form. We divide them into 2 triangles of dots,

It is easy to see that the resulting two triangles are $\triangle 3$ and $\triangle 4$, illustrating graphically that, $\triangle 4 + \triangle(4 - 1) = 4^2$

In the text we saw that the value of the words nothingness, middle, and truth is the triangle of $\triangle 37$. We then saw that the filling of these words is equal to 37^2. Since the filling of words is made up of the original letters plus additional filling letters, using the relationship just explored, we can conclude (without calculating) that the sum of just the filling letters must be $\triangle 36 = 666$!

▸ 100 is the value of the word **beauty** (יֹפִי). When we add "beauty" to "truth" (אֱמֶת), the sum is 541, or "Israel" (יִשְׂרָאֵל). Counting from 2, which is the currently accepted practice in mathematics, 541 is the 100th prime number, suggesting that it is a "beautiful" prime number, since the value of "beauty" (יֹפִי) is 100.

▸▸ In this series, the average value of 441 and 781 is 611, the value of "Torah" (תּוֹרָה) and the value of "middle" (אֶמְצָעִית) in its relational feminine form, again referring to the fact that the **goal of the Exodus** was to find this point of nothingness (that exists in the Jewish mind even when enslaved) in order to reveal the truth of the Torah at Mt. Sinai.

Torah is indeed considered the midpoint of all reality, the point that holds it all together.

▸▸▸ Extending this series backwards, we find that the number preceding 21 is 81,

81	21	61	201	441	781
-60	40	140	240	340	
	100	100	100	100	

81 is the *gematria* of the exalted form of "I" (אָנֹכִי) that begins the Ten Commandments (Exodus 20:2).

The sum of the 6 numbers from 81 to 781 is 1586 = 61 · 26, or the product of "I" (אֲנִי) times *Havayah*, alluding to the phrase frequently repeated in the Torah, "I am *Havayah*" (אֲנִי יְהוה).

Extending the series further, we find that the 7th number from 61 is 2401 = $49^2 = 7^4$. The average value of the 13 numbers, 5 before 61 and 5 after 441, is 901, the value of "the light of Mashiach" (אוֹרוֹ שֶׁל מָשִׁיחַ) and the value of the seminal blessing the Mashiach gave the Ba'al Shem Tov, "May your wellsprings spread out" (יָפוּצוּ מַעְיְנֹתֶיךָ חוּצָה).

forwards (and backwards) we use the finite differences method. Let us illustrate using our own set of 3 integers, 61, 201, and 441.

61		201		441
	140		240	
		100		

It should be evident that each number in the second row is the difference between the two numbers above it. The single number in the final row is the difference of the two differences above it and is also known as the base of the series. In this case, the base is 100.◂

Extending the series forwards and backwards we get,

21		61		201		441		781
	40		140		240		340	
		100		100		100		

So, before 61 we find 21, the square root of 441, the value of "truth." 21 is the value of God's Name, "I will be" (אֶהְיֶה), revealed to Moses at the prophecy of the burning bush. "I will be" is of course a Name referring to new birth, just as these three words, "nothingness," "middle," and "truth," refer to the blind spot in our thinking from which we are able to discover a whole new world of understanding and thought.

The next number, 781 is amazingly the value of "the middle point" (נְקוּדָה הָאֶמְצָעִית),◂◂ the exact phrase used in both the *Tanya* and *Etz Chaim* cited above to designate the middle point.[3]◂◂◂

Now that we know about the conceptual aspects of the middle point, let us turn to its mathematical manifestation. The simplest mathematical equivalent to these very lofty metaphors is the midpoint series that we have met already in chapter 2, also an example of the paradoxical nature of (countable[4]) infinity, as we saw in the previous chapter.

---

3. See also our Hebrew volume, *Mavoh Lekabbalat Ha'ari*.
4. א₀ (alef zero) in mathematical terminology.

The midpoint of Bereisheet

Let us take 913, the value of the Torah's first word, "In the beginning" (בְּרֵאשִׁית), and work our way back to its midpoint origin, which is also called its "final midpoint" ▶ (נְקוּדָה אֶמְצָעִית סוֹפִית).

$$58 < 115 < 229 < 457 < 913$$

The final midpoint of 913 is 58, which is among other things, the value of the name, Noah (נֹחַ).

Each of the five books of the Pentateuch is divided into portions, each named after one or more of the words in its opening verses. The name of the first portion (*sedra*) of the Book of Genesis is named *Bereisheet* after its first word (בְּרֵאשִׁית). As we have seen repeatedly, the value of the first word, which literally means "In the beginning" (בְּרֵאשִׁית), is 913. The name of the second Torah portion is Noah (נֹחַ). Noah literally means "rest." When his two letter name is written backwards it spells "grace" or "favor" (חֵן), as alluded to in the final verse of the portion of *Bereishit*, "And Noah found favor in the eyes of God" (וְנֹחַ מָצָא חֵן בְּעֵינֵי י-הוה).

Thus while the first word of creation, "In the beginning" (913) refers in particular to the creation of time, the final midpoint of creation (58) refers to (timeless) "rest" (נֹחַ). This timeless rest that is the final midpoint of creation is the secret of the day of Shabbat; the Shabbat is not only the seventh day of creation, it is also the day that precedes creation, which began on Sunday. Thus, the Shabbat is indeed the origin (the final midpoint) of all of creation.

Let us calculate the final midpoint of 2701, the *gematria* of the entire first verse of creation, "In the beginning, God created the heavens and the earth" (בְּרֵאשִׁית בָּרָא אֱ-לֹהִים אֵת הַשָּׁמַיִם וְאֵת הָאָרֶץ).

$$676 < 1351 < 2701$$

With only two steps, we come to the final midpoint, 676 (remember: the final midpoint is simply the even number which has no midpoint itself). $676 = 26^2$, where 26 is the value of God's essential Name

▶ The value of the phrase "**final midpoint**" (נְקוּדָה אֶמְצָעִית סוֹפִית) is 1332 = $36 \cdot 37$, also the value of the verse introducing the Ten Commandments and studied in the previous chapter.

41

Havayah (יהוה), which brings all things into existence (literally, the Name *Havayah* derives from the verb "to be"). This is indeed an amazing phenomenon.

The middle of the seven words of the first verse in Genesis is the simple two-letter word אֵת (untranslatable into English). Its numerical value is 401 (its two letters are the first and the last letters of the *alefbeit*). What is the final midpoint of this middle word?

$$26 < 51 < 101 < 201 < 401$$

Its final midpoint is 26. Again, this is the value of *Havayah* (יהוה), the square root of the final midpoint of the entire verse! The sum of the five numbers from 26 to 401 comes to 780 = △39 (39 itself is the *gematria* of "*Havayah* is one," יהוה אֶחָד). 780 = 30 · 26. The average value of the five numbers is 156 (6 · 26), the numerical value of Joseph (יוֹסֵף).

Juxtaposition of words

Finding that the final midpoint of the word אֵת, the middle word of the Torah's first verse is God's essential Name, *Havayah*, prompts us to look at the juxtaposition of these two words in the Bible's text. Having a sense of what further direction of analysis will be fruitful based on mathematical findings already discovered is one of the inner qualities that mark an individual who has been graced by God with the mission of analyzing the Torah in this way.

The first thing we notice is that in the entire Bible, the two words *et* (אֵת) and *Havayah* (יהוה) are juxtaposed 474 times. 474 is the value of "knowledge" (דַּעַת). The first appearance of the juxtaposition of the two words is in the first verse of the fourth chapter of Genesis, "Adam knew his wife Eve and she conceived and gave birth to Cain. She said, 'I have gotten a man with God'" (וְהָאָדָם יָדַע אֶת חַוָּה אִשְׁתּוֹ וַתַּהַר וַתֵּלֶד אֶת קַיִן וַתֹּאמֶר קָנִיתִי אִישׁ אֶת יהוה). The name Cain (קַיִן) is from the root of the verb here translated as "gotten" (קָנִיתִי).

Remarkably, this verse actually begins with the verb "knew" (יָדַע), the three letter root of "knowledge" (דַּעַת). The value of "knew"◄

▶ The average value of each of the letters of "knew" (יָדַע), is **28**. Half of 28 is 14, the number of words in the verse, "Adam knew his wife Eve and she conceived and gave birth to Cain. She said, 'I have gotten a man with God'" (וְהָאָדָם יָדַע אֶת חַוָּה אִשְׁתּוֹ וַתַּהַר 14 וַתֵּלֶד אֶת קַיִן וַתֹּאמֶר קָנִיתִי אִישׁ אֶת יהוה). is also the value of the first two letters of "knew" (יָדַע).

28 is also half of 56, the value of the first word of the verse, "And the man" (וְהָאָדָם). After "knew," the next two words are "et Eve" (אֶת חַוָּה), whose value is 420, another multiple of 28, 420 = 15 · 28.

(יָדַע) is 84. The average value of its three letters is 28, the triangle of 7, as we have observed above.

Out of the total 14 words in this verse, 5 are verbs. They are: "knew" (יָדַע), "she conceived" (וַתַּהַר), "she gave birth" (וַתֵּלֶד), "she said" (וַתֹּאמֶר), "I have gotten" (קָנִיתִי). The sum of their values is 2352 = 28 · 84 = 3 · 28². 2352 is also a multiple of 21, and since there are 21 letters in these 5 verbs, we find that the average value of each letter is 112, the value of God's complete Name, as found in the second account of creation, *Havayah Elokim* (יהוה אֱ־לֹהִים).

Apart from the 21 letters in the verbs, there are another 28 letters in the other words in the verse. This of course means that there are 49 = 7² letters in the entire verse. But, sectioning the verse in this way, into 21 verb-letters and 28 remaining letters allows us to note an important mathematical point regarding triangular and square numbers. 21 = △6, 28 = △7. Together, △6 ⊥ △7 = 7².

The general rule can be stated as,

> *The sum of two consecutive triangular numbers,*
> △n *and* △(n - 1) *equals the square of* n. *Or,*
>
> $$\triangle n \perp \triangle(n-1) = n^2$$

Since the entire verse contains 49 letters, let us draw it in the form of the square of 7,

ד	י	ם	ד	א	ה	ו
א	ה	ו	ח	ת	א	ע
ר	ה	ת	ו	ו	ת	ש
ק	ת	א	ד	ל	ת	ו
ר	מ	א	ת	ו	ן	י
י	א	י	ת	י	נ	ק
ה	ו	ה	י	ת	א	ש

We take this opportunity to illustrate how such a square can be analyzed.

The four corners (ו ד ש ה) equals 315 = 7 · 45, where 45 is the value of "man," or "Adam" (אָדָם). Let us extend the corners a bit,

ד	י	ם	ד	א	ה	ו
א	ה	ו	ח	ת	א	ע
ר	ה	ת	ו	ו	ת	ש
ק	ת	א	ד	ל	ת	ו
ר	מ	א	ת	ו	ן	י
י	א	י	ת	י	נ	ק
ה	ו	ה	י	ת	א	ש

The *gematria* of the 12 letters not dimmed (והע ידא קשא יוה) is 518 = 14 · 37, where 37 is the value of "Abel" (הֶבֶל). The remaining 37 (dimmed) letters of the square come to 4316 = 26 · 166. 26 is of course the value of *Havayah*, God's essential Name. 166 is the value of one of God's mystical Names in Kabbalah. ◄

The square's middle column is,

ד	י	ם	**ד**	א	ה	ו
א	ה	ו	**ח**	ת	א	ע
ר	ה	ת	**ו**	ו	ת	ש
ק	ת	א	**ד**	ל	ת	ו
ר	מ	א	**ת**	ו	ן	י
י	א	י	**ת**	י	נ	ק
ה	ו	ה	**י**	ת	א	ש

The value of the letters in the middle column (ד ח ו ד ת ת י) is 832, also a multiple of 26, in this case 26 · 32. 832 is also the *gematria* of the "Land of Israel" (אֶרֶץ יִשְׂרָאֵל).

► One of the 4 primary fillings of *Havayah* is יוד הי ואו הי, whose value is 63. This filling is referred to as the Name *sag* because *sag* is a phonetization of its value, 63 (סג). Writing this filling in its generative form gives, יוד יוד הי יוד הי ואו יוד הי ואו הי whose value is 166, considered **the mystical Name 166**.

Apart from the root letters of *Havayah* in this Name, the value of the additional filling letters (יוד הי ואו הי) is 37.

The final column is,

```
ד   י   ם   ד   א   ה   ו
א   ה   ו   ח   ת   א   ע
ר   ה   ת   ו   ו   ת   ש
ק   ת   א   ד   ל   ת   ו
ר   מ   א   ת   ו   ן   י
י   א   י   ת   י   נ   ק
ה   ו   ה   י   ת   א   ש
```

The value of its letters (דארקריה) is 520, another multiple of 26: 520 = 20 · 26.

The sum of the middle and last columns is then, 832 ⊥ 520 = 1352 and their average value is then 1352/2 = 676, or 26^2!

The inner 5 by 5 square is,

```
ד   י   ם   ד   א   ה   ו
א   ה   ו   ח   ת   א   ע
ר   ה   ת   ו   ו   ת   ש
ק   ת   א   ד   ל   ת   ו
ר   מ   א   ת   ו   ן   י
י   א   י   ת   י   נ   ק
ה   ו   ה   י   ת   א   ש
```

Its letters come to 3040, the product of 19 and 160. 19 is the value of "Eve" (חַוָּה) and 160 the value of Cain (קַיִן), the two subjects of the verse!

Finally, the square's external border is,

ד	י	מ	ד	א	ה	ו
א	ה	ו	ח	ת	א	ע
ר	ה	ת	ו	ו	ת	ש
ק	ת	א	ד	ל	ת	ו
ר	מ	א	ת	ו	ו	י
י	א	י	ת	י	נ	ק
ה	ו	ה	י	ת	א	ש

Its letters equal 1794, another multiple of 26: 1794 = 26 · 69, or the product of 78 · 23, where 23 is the value of Eve's original name (before she sinned), *Chayah* (חַיָּה).

God's image

Continuing with our midpoint series analysis of the Torah's first verse, let us look at the final midpoints of the second and third words. The second word is "created" (בָּרָא) and its value is 203. Its midpoint series is thus,

$$102 < 203$$

The third word is *Elokim* (אֱ־לֹהִים), whose value is 86. By definition, an even number is its own midpoint, so we conclude that 86 is its own final midpoint.

The sum then of the final midpoints of the first three words of the Torah, "In the beginning God created" (בְּרֵאשִׁית בָּרָא אֱ־לֹהִים)—58, 102, 86—is 246, the value of "image of God" (צֶלֶם אֱלֹהִים) in which man was created.

Man is the culmination and pinnacle of creation, and so we see that "the end [of creation] is enwedged into the beginning [of creation]."[5] The *gematria* of the word "image" (צֶלֶם) is 160 = 58 + 102, the final midpoints of the first two words of the Torah, "In the beginning" (בְּרֵאשִׁית) and "created" (בָּרָא), respectively. 58 is also the *gematria*

---

5. *Sefer Yetzirah* 1:7.

of "grace" (חֵן), as we saw earlier, and 102 is the *gematria* of "faith" (אֱמוּנָה). From this mathematical equality, we discover that the secret of the "image [of God]" (צֶלֶם) is the same as "grace" of "faith [in God]" (חֵן אֱמוּנָה). Furthermore, note that the Name of "God" used in the idiom "image of God" is God's Name *Elokim* (אֱ־לֹהִים), the Torah's third word, whose final midpoint is equal to itself (since its *gematria*, 86, is an even number).

We saw above that

$$10 < 19 < 37 < 73$$

Let us continue this progression of successive midpoints to seven places (which will place 73 in the middle):

$$10 < 19 < 37 < 73 < 145 < 289 < 577 \blacktriangleright$$

The first word of the Torah, "In the beginning" (בְּרֵאשִׁית), alludes to wisdom, as in the verse, "The beginning of wisdom…."▶▶ The *gematria* of wisdom, as we saw, is 73, its ordinal value is 37 and its reduced value is 19.

The next two words of the Torah, "God created" (בָּרָא אֱ־לֹהִים) equal 289, the sixth number of the above progression (defined by the three values of wisdom, 19, 37, 73). 289 is also 17^2, where 17 is the value of "good" (טוֹב). The sum of these first 7 numbers of the series is 1150 = 50 · 23; 23 is the companion number of 37.

The average value of the 3 succeeding numbers in the progression, 73, 145, 289, from "wisdom" (חָכְמָה) to "God created" (בָּרָא אֱ־לֹהִים), is $169 = 13^2$

The average value of just 73 and 289 is 181, the inspirational number of 10, i.e., $⊡10 = 181 = 10^2 ⊥ 9^2$.

> *We use the notation ⊡n to denote the inspirational, or interface number of n defined as,*
> $$⊡n = n^2 ⊥ (n - 1)^2$$

10 is of course the final midpoint that begins our series.

181 is the midpoint of $361 = 19^2$.▶▶▶

▶ Note that 577 is the decimal adjusted form of Euler's constant, 0.577…

▶▶ Grammatically, "In the beginning" (בְּרֵאשִׁית) is in the construct form (סְמִיכוּת), requiring that another word follow it, a word that is seemingly missing from the text (see *Rashi's* commentary to the verse). Since "In the beginning" alludes to wisdom, as in, "**the beginning of wisdom**" and since the *gematria* of the entire verse is the triangle of 73, wisdom, we may contemplate what happens if the seemingly missing word is indeed "wisdom." [Of course no word is actually missing, as the Torah is complete].

The reading of the verse would become, "With the beginning of wisdom [the fear of *Havayah*], God created the heavens and the earth" (בְּרֵאשִׁית חָכְמָה בָּרָא אֱ־לֹהִים אֵת הַשָּׁמַיִם וְאֵת הָאָרֶץ). In this hypothetical version, the verse contains 32 letters, corresponding to the 32 pathways of wisdom with which God created the world, as taught in the beginning of *Sefer Yetzirah*.

Further analysis yields that, "With the beginning of wisdom" (בְּרֵאשִׁית חָכְמָה) is 986 = 17 · 58, where 17 is "good" (טוֹב) and 58 is "grace" (חֵן). The first four words are then, "With the beginning of wisdom God created" and their value is 1275 = 17 · 75 = △50. The *gematria* of all 8 words of the hypothetical reading are then 2774 = 38 · 73, wisdom (the triangle of wisdom and wisdom itself).

▶▶▶ The algebraic rule that can be extracted from this relationship is, that since the definition of the interface value of n, ⊡n = $n^2 ⊥ (n - 1)^2$, then,

$$⊡n < (2n - 1)^2$$

where the right hand side of the equality is simply the square of the sum of the two roots that make up the interface number.

47

▶ Wisdom and beauty are related mathematically. Taking the triangular value of each letter in "wisdom" (חָכְמָה), we find that their sum is $\triangle 8 + \triangle 20 + \triangle 40 + \triangle 5 = 36 + 210 + 820 + 15 = 1081$, the gematria of "beauty" (תִּפְאֶרֶת).

▶▶ As noted earlier (p. 33), every even number seeds an infinite **midpoint series**. Let us look at the sum of the first n entries in the midpoint series beginning with the even numbers n = 2, 4, 6, 8, and 10.

n	first n entries	sum
2	2, 3	5
4	4, 7, 13, 25	49
6	6, 11, 21, 41, 81, 161	321
8	8, 15, 29, 57, 113, 225, 449, 897	1793
10	10, 19, 37, 73, 145, 289, 577, 1153,	9217
	2305, 4609	

Now, the sum of the sums is 11385, which is the product of 45 and 253, both triangular numbers. Specifically, 45 is the triangle of 9 and 253 is the triangle of 22. These two numbers 9 and 22 allude to a basic structure of the Hebrew alphabet: 22 is of course the number of letters in the alphabet, while 9 is the number of "pronunciation marks" (nikud).

If we continue the "wisdom" midpoint series, the next number will be 1153. We then find that the average value of the 4 consecutive numbers that follow "wisdom" (חָכְמָה) 73, 145, 289, 577, and 1153 is 541, the value of "Israel" (יִשְׂרָאֵל), the last word of the Torah that according to the rule that, "the end is enwedged in the beginning," connects back to the first word, "In the beginning."

The average value of the 4 numbers beginning with "wisdom," 73, is 271, the midpoint of 541. And the average value of the 4 numbers beginning with 37 is 136, the midpoint of 271.

Following 1153 comes 2305. The average value of the 4 numbers concluding with 2305 is 1081, the gematria of "beauty" (תִּפְאֶרֶת), the sixth of the Divine sefirot, the central sefirah of the emotive attributes of the soul. Thus wisdom, the first of the conscious sefirot, generates through its series of midpoints beauty, the sixth sefirah.

In Kabbalah, wisdom is the father and beauty is the son◀ and at the same time 1081, the value of "beauty" (תִּפְאֶרֶת) is equal to $\triangle 46$. 541, "Israel" (יִשְׂרָאֵל), is the midpoint of 1081.

The sum of the first 9 numbers of the series, 10, 19, 37, 73, 145, 289, 577, 1153, and 2305 is 4608◀◀ $= 2 \cdot 48^2$, also called the double square of 48, where 48 is the value of "brain" (מֹחַ). The double square indicates that this number is connected with both brain lobes, representing the two intellectual sefirot wisdom and understanding, the father and mother principles that give birth to the 7 emotions of the heart, corresponding to the 7 days of creation.

Midpoint series and the square numbers

As already stated, every even number begins an infinite midpoint progression of odd numbers. What we now note is that the midpoint series beginning with 10 quickly "converges" to a perfect square, $289 = 17^2$. Let us look at this phenomenon in respect to other midpoint series.

The first midpoint series, beginning with 2, is

$$2 < 3 < 5 < 9$$

It even more quickly converges to $9 = 3^2$.

The next midpoint series is,

$$4 \lessdot 7 \lessdot 13 \lessdot 25 \lessdot 49$$

It converges just as quickly to $25 = 5^2$ and then we have another square number $49 = 7^2$▶

The next midpoint series, begins with 6,

$$6 \lessdot 11 \lessdot 21 \lessdot 41 \lessdot 81$$

We see that it converges to $81 = 9^2$.

The next midpoint series converges to $225 = 15^2$,

$$8 \lessdot 15 \lessdot 29 \lessdot 57 \lessdot 113 \lessdot 225$$

One might by now be convinced that quick convergence to a perfect square is an intrinsic property of all midpoint progressions. But in mathematics (and certainly in secrets of creation), one should never jump to quick conclusions (without rigorous proof). Examine the next midpoint series (after the one starting from 10, with which we began):

$$12 \lessdot 23 \lessdot 45 \lessdot 89 \lessdot 177 \lessdot 353 \lessdot 705 \lessdot 1409 \lessdot 2817 \lessdot 5633\ldots$$

It's going to take us quite a while until we reach a perfect square, if at all. The same is true for the next midpoint series, beginning with 14,

$$14 \lessdot 27 \lessdot 53 \lessdot 105 \lessdot 209 \lessdot 417 \lessdot 833 \lessdot 1665 \lessdot 3329\ldots$$

What about the next progression?

$$16 \lessdot 31 \lessdot 61 \lessdot 121 \lessdot 241 \lessdot 481 \lessdot 961$$

Three steps from 16 gives $121 = 11^2$, and another three steps from 121 gives $961 = 31^2$.▶▶

So far we have found the squares of 3, 5, 7, 9, 11, 15, 17, and 31. What about the square of 13 (=169). To reach 13^2, we must turn to the midpoint series originating from 22,

$$22 \lessdot 43 \lessdot 85 \lessdot 169$$

To reach the square of 19, we must begin with 46.

▶ Considering this midpoint series,

$$4 \lessdot 7 \lessdot 13 \lessdot 25 \lessdot 49$$

Since, $49 = 7^2$ we find that the square of 5 is the midpoint of the square of 7. This is a relatively rare phenomenon. If we would like to better understand it, we have to look at the next case in which a square number is the midpoint of another square. The next pair of squares for which this occurs is,

$$841 \lessdot 1681$$

where $841 = 29^2$ and $1681 = 41^2$. 29^2 is the midpoint of 41^2.

Common to both 25 and 841 is that both are **numbers that are both squares and interface numbers**. $25 = 3^2 \perp 4^2$ and $841 = 20^2 \perp 21^2$.

Earlier, we obtained a result that interface numbers are always the midpoint of the square obtained from the sum the two roots making up the interface number. Or,

$$\square n \lessdot (2n - 1)^2$$

So, we may conclude that when a square number is also an interface number, it will itself be the midpoint of a square.

▶▶ The average value of these two squares, 11^2 and 31^2, is 541, the *gematria* of "Israel" (יִשְׂרָאֵל). The difference between 541 and 121 is 420, as is the difference between 961 and 541. 420 is the *gematria* of the union of Jacob and Rachel (יַעֲקֹב רָחֵל).

420 = ♠20 (note that △20 = 210 = ♠14).

49

21^2 is generated from 56, 23^2 by 34, 25^2 by 40, 27^2 by 92, 29^2 by 106.

There does not seem to be an easily definable rule determining in which midpoint series each square number will appear. Nonetheless, we definitely see that quick convergence to a perfect square is an outstanding property of the midpoint series from 2 up to and including 10 and then with the series beginning with 16.◀ These two numbers, 10 and 16 denote the division of *Havayah* into its first letter, *yud* (י), which equals 10, and to its remaining three letters, *hei vav hei* (הוה), whose value is 16. Since each of the letters corresponds to one of the Four Worlds, this division parallels the essential separation between the highest World, Emanation, and the three lower Worlds, Creation, Formation, and Action.

As its name suggests, a perfect square is a sign of perfection, or, in the terminology of Kabbalah and Chasidut, of consummate inter-inclusion (each root element reflecting all of the elements of the root set). In Kabbalah, a perfect square symbolizes the complete "rectification" (תִּקּוּן) of created reality, the state of reality that God intended in creation. This is indicated by the fact that the *gematria* of, "God created" (בְּרָא אֱ־לֹהִים) is 289 = 17^2, where 17 is the value of "good" (טוֹב), so all that God has created is in essence "good squared." We have already seen that this is the square generated by the "wisdom" midpoint series beginning with 10, the number of the *sefirot* (Divine emanations, channels of creative energy) with which God created the world.

Midpoint of the fine structure constant

We will conclude this section with another important midpoint phenomenon. The number 137—the *gematria* of Kabbalah (קַבָּלָה)—is perhaps the most important pure number in modern physics. It is the inverse of the fine structure constant. Consider its midpoint series,

$$18 < 35 < 69 < 137 < 273 < 545 < 1089$$

▸ The 10 and 16 midpoint series converge to 17^2 and 31^2 respectively. The average value of these two squares is 625 = 25^2 = 5^4. 5 is the value of the letter *hei* (ה) and the sages teach us that the world was created with the letter *hei* (*Bereisheet Rabah* 12:10 and *Rashi* to Genesis 2:4).

The series begins with 18, the value of "alive" (חי). The 7th number in this midpoint series is 1089 = 33^2.► The average value of 137 and 1089 is 613, the number of commandments in the Torah and the 18th inspirational number, i.e., 18^2 ⊥ 17^2. The sum of the first four numbers of the series, from 18 to 137, is 259 = 7 · 37, where 7 alludes to the 7 vapors of creation and the 7 vanities of Ecclesiastes, a topic we will explore in greater depth later on. The third number, 69 = 3 · 23, the companion number of 37, as already explained. The sum of the first 7 numbers in this midpoint series is 2166 = 6 · 361, or 19^2.

► Note that if we take 2 to be the first prime number (as is the modern mathematical convention), 137 is the 33rd prime number.

LEVELS OF GEMATRIA

We'll begin our discussion of the various levels of *gematria* with another observation about the Torah's first verse. The *gematria* of the Torah's first verse, "In the beginning God created the heavens and the earth" (בְּרֵאשִׁית בָּרָא אֱ-לֹהִים אֵת הַשָּׁמַיִם וְאֵת הָאָרֶץ) is 2701, which as noted in the previous chapters is a multiple of 37, specifically: 2701 = 37 · 73. We have also noted that the numerical values of each of the two final words in the verse, "and the earth" (וְאֵת הָאָרֶץ), which are 407 and 296, respectively, are both multiples of 37.

This leads us to ask how many multiples of 37 can be obtained by summing the words of the first verse in various combinations. Altogether 7 words can be combined in 128 different ways▸. Of these, let us write only those combinations whose sum is a multiple of 37. For brevity, we will write these out in Hebrew only:

1. בְּרֵאשִׁית אֱ-לֹהִים = 999 = 37 · 27
2. בְּרֵאשִׁית בָּרָא אֵת = 1517 = 37 · 41
3. בְּרֵאשִׁית אֱ-לֹהִים וְאֵת = 1406 = 37 · 38
4. בְּרֵאשִׁית אֱ-לֹהִים הָאָרֶץ = 1295 = 37 · 35
5. בְּרֵאשִׁית אֱ-לֹהִים וְאֵת הָאָרֶץ = 1702 = 37 · 46
6. בְּרֵאשִׁית בָּרָא אֵת וְאֵת = 1924 = 37 · 52
7. בְּרֵאשִׁית בָּרָא אֵת הָאָרֶץ = 1813 = 37 · 49
8. בְּרֵאשִׁית בָּרָא אֱ-לֹהִים אֵת הַשָּׁמַיִם = 1998 = 37 · 54
9. בְּרֵאשִׁית בָּרָא אֵת וְאֵת הָאָרֶץ = 2220 = 37 · 60
10. בְּרֵאשִׁית בָּרָא אֱ-לֹהִים אֵת הַשָּׁמַיִם וְאֵת = 2405 = 37 · 65
11. בְּרֵאשִׁית בָּרָא אֱ-לֹהִים אֵת הַשָּׁמַיִם הָאָרֶץ = 2294 = 37 · 62
12. בְּרֵאשִׁית בָּרָא אֱ-לֹהִים אֵת הַשָּׁמַיִם וְאֵת הָאָרֶץ = 2701 = 37 · 73
13. בָּרָא אֵת הַשָּׁמַיִם = 999 = 37 · 27
14. בָּרָא אֵת הַשָּׁמַיִם וְאֵת = 1406 = 37 · 38

▸ From combinatorics, we get the result that 7 words can be combined in 128 different ways (1 + 7 + 21 + 35 + 35 + 21 + 7 + 1), but of these one is actually the empty set with 0 words (choosing 0 out of 7). Therefore, realistically, there are only 127 ways to combine 7 words.

The number 127 is tied to the Matriarch Sarah's lifespan (Genesis 23:1). 127 is also a prime number whose triangle is a perfect number.

Note that both 37 and 127 are Shabbat numbers, numbers that are generated by the function

$$f[n] = 6 \triangle n + 1$$

The first four Shabbat numbers are 1, 7, 19, 37 and they have the following form:

37 is thus the 4th Shabbat number while 127 is the 7th. The relationship between 4 and 7 is that 4 is the midpoint of 7.

15. $35 \cdot 37 = 1295 =$ בָּרָא אֵת הַשָּׁמַיִם הָאָרֶץ

16. $46 \cdot 37 = 1702 =$ בָּרָא אֵת הַשָּׁמַיִם וְאֵת הָאָרֶץ

17. $13 \cdot 37 = 481 =$ אֱ־לֹהִים הַשָּׁמַיִם

18. $24 \cdot 37 = 888 =$ אֱ־לֹהִים הַשָּׁמַיִם וְאֵת

19. $21 \cdot 37 = 777 =$ אֱ־לֹהִים הַשָּׁמַיִם הָאָרֶץ

20. $32 \cdot 37 = 1184 =$ אֱ־לֹהִים הַשָּׁמַיִם וְאֵת הָאָרֶץ

21. $11 \cdot 37 = 407 =$ וְאֵת

22. $19 \cdot 37 = 703 =$ וְאֵת הָאָרֶץ

23. $8 \cdot 37 = 296 =$ הָאָרֶץ

▸ The sum of all the multipliers of 37 in these 23 different combinations is 876, which is the product of 12 and 73, where 73 is the value of the largest multiplier. Note also that 37 is the midpoint of 73 and if we add 37 to 876, we get 913, the value of "In the beginning" (בְּרֵאשִׁית)!

Here is another interesting relationship between **23 and 37**. Counting 1 as the first prime number 23 is the 10th prime and 37 is the 13th, making the sum of their ordinals 23.

▸▸ The product of 23 and 37 is 851. The *gematria* of the Torah's first verse is 2701. Subtracting 851 from 2701 leaves 1850, also a multiple of 37, specifically 1850 = 50 · 37.

But, note that when the 1000 in 1850 is replaced by a 1 (as we've seen on p. 31), 1850 reduces to 851, once again, the product of 23 and 37!

We have identified 23 independent combinations◂ of words in the first verse of Genesis whose sum is a multiple of 37!◂◂ We have already noted and discussed in previous chapters how the numbers 23 and 37 are companion numbers commonly appearing together throughout the Torah in a plethora of various issues.

One beautiful and simple illustration of the relationship between 23 and 37 in the Torah's first verse is the following. When the numerical value of each word in the first verse is reduced to one digit, their values become: 4, 5, 5, 5, 8, 2, and 8; these seven numbers themselves sum to 37. However, if we look at the one-digit values of just the four words that make up the second half of the verse, "the heavens and the earth" they are 5, 8, 2, and 8, whose sum is 23.

Different systems of *gematria*

Let us now deepen our understanding of the reduced single-digit *gematria* value and its relationship with the other possible *gematria* values. Above we saw that the normative *gematria* value of "wisdom" (חָכְמָה) is 73, but that in ordinal *gematria* it equals 37, and its reduced value is 19. Note that the value of "wisdom" in each system of *gematria*, whether it is 73, 37, or 19, can ultimately be reduced to 1 by repeatedly adding the digits together until reaching a single digit from 1 to 9.

1 is thus the final reduced value of wisdom at all of its levels. Wisdom is indeed the first (1) conscious faculty of the soul. ▶ It is the beginning of creation, the Divine power by which God created the world.[1]

A beautiful result that can be proven mathematically is that all three *gematria* values (normative *gematria*, ordinal *gematria*, and reduced numbering) of any word or phrase, will ultimately reduce to the same number from 1 to 9 each. To offer another illustration of this point, let us look at the Torah's first verse. Its normative value is 2701—adding the digits together we get 10, which further reduces to 1. ▶▶ The ordinal value of the Torah's first verse is 298—adding the digits together yields 19, which further reduces to 10, and then to 1. Finally, the reduced numbering value of the Torah's first verse is 82, which again reduces to 10, and then to 1. ▶▶▶

The light in the letters

What is the origin and validation for these different systems of *gematria*, or stated more exactly, for these different numbering schemes used in Kabbalah. To answer this question, let us offer some words of introduction.

Every word in any language is meant to convey some meaning. This is true to the greatest extent in Hebrew, the language with which the Almighty created reality in all its dimensions. Meaning is carried by the letters of the word. The basis of *gematria* is that every letter in the Hebrew alphabet has a numerical value and thus, every word has a numerical value. But, the numerical value of a word in Hebrew, i.e., its *gematria*, refers to the lowest level of energy or light carried by its letters. Light in this context refers to the individuality level exhibited by the word when analyzed in this particular manner. This is the reason that we find that when analyzed numerically, there are many words that equal one another. It is of course not true that anything can be shown to equal anything, but it is true that *not* every numerical equality between words is meaningful. To properly

---

1. *Targum Yerushalmi* on Genesis 1:1.

▶ In the *Tanya* (ch. 35) it is explained, in the name of the Magid of Mezritch, that the "True One" (1) manifests only in the *sefirah* of wisdom.

▶▶ 2701 reduces to 10 and then to 1. Note that we can obtain the same result by reducing the values of the multiplicands whose product is 2701. In this case they are 37 and 73 and reducing them we find that 37 goes to 10 and then to 1 and likewise, 73 goes to 10 and then to 1, so their product is 1!

If we try to generalize this, we will find that this is the case only for **numbers that are the product of exactly 2 primes**. For example, 33 is the product of 3 and 11, both primes. 33 reduces to 6, while 3 remains 3 and 11 reduces to 2 and their product is also 6.

▶▶▶ The sum of the normative, ordinal and reduced values of the Torah's first verse is: 2701 + 298 + 82 = 3081, itself a triangular number, 3081 = △78.

explain the meaning of a numerical equality requires the most advanced Torah scholarship. Therefore, only equalities explained by the great Torah scholars throughout the generations are considered to carry significant meaning.[2]

All this when discussing the normative numerical value of a word. Obviously, when the normative numerical value is reduced, as it is in reduced value *gematria*, the light carried by the number reduces even further, decreasing the differentiation, or individuality between different words. The greater the reduction, the less the differentiation, until when reaching the final reduced value, which can be only one digit from 1 to 9, the chances of a word or concept being equal to another increases to 1 in 9.

Practically speaking, this means that the fact the normative value of "wisdom" (חָכְמָה) is 73 is more meaningful than the fact that its ordinal value is 37. Its reduced value (19) carries even less differentiating energy and its final reduced value (1) is the least meaningful and differentiating of all.

Levels of reduction

So far we have referred to the levels of meaning associated with the various forms of *gematria* in general terms. Let us be more specific. In Kabbalah, the 4 levels of numerical value that are organized according to levels of light are normative *gematria*, ordinal *gematria*, reduced numbering, and final reduced value. These four levels correspond in order to the Four Worlds: Emanation (אֲצִילוּת), Creation (בְּרִיאָה), Formation (יְצִירָה), and Action (עֲשִׂיָּה).

type of *gematria*	values	World
normative (הָכְרֵחִי)	1…10, 10…100, 100…400	Emanation
ordinal (סִדּוּרִי)	1…22	Creation
reduced (קָטָן)	1…9 cumulative	Formation
final reduced (קָטָן אַחֲרוֹן)	1…9	Action

---

2. See *Likutei Sichot* v. 26, pp. 206ff.

Normative numbering and Emanation

In the two higher worlds of Emanation and Creation, each of the 22 letters of the *Alef-beit* possesses a unique numerical value, a property that clearly differentiates them from the numbering schemes found in the two lower worlds, where more than one letter can share the same value.

The world of Emanation has another unique property that sets it apart from the numbering scheme found in the World of Creation. Emanation is a world of pure Divine consciousness, the consciousness of the supernal *sefirot* that channel Divine light and energy into created reality. There, the values of the letters are based on normative numbering following a one-to-one mapping of letters to values, but with quantum leaps. Let us explain. Though the reader who has reached this stage of the book is obviously familiar with normative value, in the interest of clarity we include the mapping of letters to values in this numbering scheme:

letter	name	value	letter	name	value	letter	name	value
א	alef	1	י	yud	10	ק	kuf	100
ב	bet	2	כ	kaf	20	ר	reish	200
ג	gimel	3	ל	lamed	30	ש	shin	300
ד	dalet	4	מ	mem	40	ת	tav	400
ה	hei	5	נ	nun	50			
ו	vav	6	ס	samech	60			
ז	zayin	7	ע	ayin	70			
ח	chet	8	פ	pei	80			
ט	tet	9	צ	tzadik	90			

Notice that the normative numbering found in the World of Emanation experiences quantum leaps when assigning values to the letters. When reaching 10 (the letter *yud*) a quantum leap takes place and instead of continuing with 11 the 10 itself bounces to a new energy level paralleling the first and continues from 10 to 100 (10, 20,

30,…). Similarly, when the value reaches 100 (the value of the letter *kuf*), it once again skips to a new energy level (100, 200, 300, 400).

Ordinal numbering and Creation

Clearly, the quantum leap in value found in the World of Emanation leads to greater differentiation or distance[3] between words when their value is calculated using normative numbering than when it is calculated using the ordinal numbering of the World of Creation, whose values are sequential, with no leaps in between. Indeed, the ordinal, sequential numbering of the Creation perfectly fits its nature as the world of logical and linear order. Here it makes best sense to assign each of the 22 letters a corresponding value from 1 to 22.

Reduced numbering and Formation

The differentiation or distance between words lessens as we descend into the World of Formation, where the value of a word is calculated as the sum of the values of its letters. But, the letters themselves do not have unique values. Instead, the values are limited to from 1 to 9, as follows,

letter	name	value	letter	name	value	letter	name	value
א	alef	1	י	yud	1	ק	kuf	1
ב	bet	2	כ	kaf	2	ר	reish	2
ג	gimel	3	ל	lamed	3	ש	shin	3
ד	dalet	4	מ	mem	4	ת	tav	4
ה	hei	5	נ	nun	5			
ו	vav	6	ס	samech	6			
ז	zayin	7	ע	ayin	7			
ח	chet	8	פ	pei	8			
ט	tet	9	צ	tzadik	9			

---

3. Similar in sense to the mathematical measure of distance between two words used in information theory and referred to as Hacking distance.

That reduced numbering corresponds to the World of Formation is learnt from several places in the *Zohar*,[4] where this numbering system is referred to as the "reduced numbering of Chanoch [i.e., the Biblical Enoch]." In Kabbalah, Chanoch is referred to as a "youth" (נַעַר) that inhabits the World of Formation. ▸ Indeed, young children (or adults who are still at a certain level of immaturity) are only capable of (or possess an affinity to) calculating the letters of the *Alef-beit* according to their small numbering value (each letter from 1 to 9). From the correspondence between reduced numbering and the World of Formation (stated explicitly in the Zohar), it is possible to correctly draw the correspondence between the other 3 types of numbering and their respective Worlds, as we have done here.

Final reduced numbering and Action

Further loss in differentiation between words is experienced once we descend into the World of Action, where the individual values of letters are gone altogether and the value of a word is calculated based on the digits of its reduced value alone; only the final reduction of the sum of the letters remains to characterize the word.

This almost complete loss of differentiation fits well with Isaiah's portrayal of the World of Action being literally severed from the three higher worlds of Emanation, Creation, and Formation. His use of the word "even" (אַף) in the verse, "… I have created, I have formed, even have I made"[5] implies that the World of Action suffers, as it were, from an existential leap in the level of Divine concealment. Whereas the Divine is clearly present in each of the higher three worlds, externally, God seems to have completely disappeared from the fourth—the World of Action. Just as in regard to *gematria*, the particular letters of the Hebrew words from which all of reality comes into being, cannot be resolved in final reduced numbering, so in the World of Action only natural laws remain with the Creator's

▸ One of the instances when the word "youth" appears in a verse is "Joseph was seventeen years old… and he was a youth" (Genesis 37:2). Since, the word "youth" (נַעַר) refers to Enoch and the angel *Metat* who dwell in the World of Formation, the *Shelah* (*parashat Vayeishev s.v. uve'inyan*) explains that the number 17 in the verse actually alludes to the reduced value of God's essential Name, *Havayah*, which is 17. In reduced numbering *yud* equals 1 and the rest of the letters equal 16, just as they would in normative number.

---

4. III, 253a. See also *Pardes Rimonim*, 30:8.
5. Isaiah 43:7.

Presence almost indeterminable from them. Just as in the case of reduced numbering corresponding with the World of Formation, this very clear correspondence between final reduced numbering and the World of Action can serve as a basis for extrapolating the parallels between the 3 higher *gematria* systems and their respective Worlds.

Now that we have introduced the various numbering systems in Kabbalah and seen how they correspond to the Four Worlds, we can look at more in-depth spiritual relationships hidden in the Torah.

WORLDS INTERRELATE

Now that we have established how the four major numbering schemes correspond to the Four Worlds, it will be revealing to look at two types of mathematical relationships. First, we will look at instances where two *different* words have the same value in two *different* numbering schemes. Then, we will look at examples of the relationships between different values for the *same* word in *different* numbering schemes. In both cases, we will of course focus on the Torah's first verse.

When worlds connect

Our first example returns to the word "wisdom." In the previous chapters, we have seen how important wisdom is in respect to creation and particularly the first verse of the Torah's account of creation.

In ordinal numbering, i.e., the numbering system of the World of Creation, the value of "wisdom" (חָכְמָה) is 37. 37 is also the normative value of "vanity" (הֶבֶל), the numbering system of Emanation, whereas "vanity" in ordinal numbering—indicating numerical value in the World of Creation—is 19. But, 19 is itself the value of "wisdom" in reduced numbering—the numbering system corresponding to the World of Formation.

This is not a mathematically necessary relationship for any arbitrary word. Just because the value of word A in one World equals that of another word, B, in a lower World, does not necessitate that A's value in that lower World will equal the value of B in a still-lower World.

Let, us state this more rigorously. Let us use the subscript letters, e, c, f, and a to denote the Four Worlds (Emanation, Creation,

Formation, and Action). Thus A_e denotes the value of word A in Emanation, i.e., in normative *gematria*. Using this notation we can state that:

$$A_e = B_c \text{ does not imply } A_c = B_f$$

And more generally, substituting the ordered set n = {4, 3, 2, 1} for the Four Worlds (from Emanation to Action), we can write that,

$$A_n = B_{n-1} \text{ does not imply } A_{n-1} = B_{n-2}$$

What all this reveals is that the numerical relationship between wisdom and vanity indeed holds meaning worthy of our attention.

Indeed, in the second verse of Ecclesiastes◄, King Solomon, the wisest of men, states that, "Vanity of vanities says Ecclesiastes, vanity of vanities, all is vanity."◄◄

The sages[1] explain that this verse alludes to the world having been created with 7 vanities◄◄◄. Thus, reality was created both with wisdom and with vanity, or put more succinctly: a world created with wisdom is a world of vanity.

Now, why wisdom gives rise to vanity is a good question, but it seems that you have to be the wisest of men to realize this consequence. We will revisit the connection between vanity and wisdom in the next chapter. In the meantime, let us take this opportunity to see another beautiful relationship between Ecclesiastes and the first verse of Genesis.

Even though the verse proclaiming that all is vanity is the second verse of the Book of Ecclesiastes, it actually follows an introductory verse and so to a certain extent can be considered the first verse of content. This is similar to the straightforward understanding that though no verse precedes the Torah's first verse, in effect the Torah is the word of God and therefore, it is as if there is an introductory verse that states that, "These are the words of God spoken to Moses," preceding the first verse. But, the connection between the second

► **Ecclesiastes**, or *Kohelet* in Hebrew, literally means, "one who congregates people to teach them."

►► The first verse in Ecclesiastes reads, "The words of Ecclesiastes the son of David, King in Jerusalem" (דִּבְרֵי קֹהֶלֶת בֶּן דָּוִד מֶלֶךְ בִּירוּשָׁלָם). The numerical value of this first verse is 1495, the combined **value of all 22 letters** of the Hebrew alphabet, אבגדהוזחטיכלמנסעפצקרשת.

►►► How do the sages count **the 7 vanities** in this verse? Following the linguistic rule that "the plural form of a word refers to the minimum plurality", or 2, each plural form "vanities" refers to two vanities, and so "vanity of vanities" refers to three vanities. Since this phrase repeats twice, we have six vanities. The final phrase, "all is vanity" increases the count to 7 vanities.

1. *Kohelet Rabbah* 1:3.

verse of Ecclesiastes and the Torah's first verse runs far deeper. The *gematria* of "Vanity of vanities says Ecclesiastes, vanity of vanities, all is vanity" (הֲבֵל הֲבָלִים אָמַר קֹהֶלֶת הֲבֵל הֲבָלִים הַכֹּל הָבֶל) is 1116. But, 1116 is also the *gematria* of the first two words of the Torah, "In the beginning [God] created" (בְּרֵאשִׁית בָּרָא).

1116 is also the *gematria* of the phrase "crown of kingdom" (כֶּתֶר מַלְכוּת), a phrase that appears three times in the book of Esther.[2]► The first two words of the Torah are the Creator's "crown of kingdom," the crown of His sovereignty over His kingdom—all that He has created.

Creation with compassion and judgment

The ordinal *gematria* of the Torah's first verse is 298, which in normative *gematria* is the value of "mercy" or "compassion" (רַחֲמִים). At the same time the ordinal value of just the two words, "God created" (בָּרָא אֱלֹהִים) is 64, which is the value of "judgment" (דִין) in normative *gematria*.►►

How do these two relationships play out? The sages teach that God's original (as it were) intent was to create the world with His Divine attribute of judgment (מִדַּת הַדִּין), but realizing that the world would be unsustainable in the face of His judgment, He introduced into the Working of Creation His Divine attribute of compassion (מִדַּת הָרַחֲמִים).[3]►►► This teaching is most beautifully alluded to in the above numerical phenomenon, where even though the ordinal *gematria* of the complete verse reveals God's attribute of compassion—sustaining creation—hidden deeper, within just the two words that describe the actual act of creation lies His original intent of creating the world with judgment.

In the previous chapter, we noted that the World of Emanation and normative numbering include quantum leaps that are above and beyond the rational logic of the World of Creation and ordinal

---

2. 1:11, 2:17, 5:8.
3. See *Rashi* to Genesis 1:1.

► The first time the phrase "**crown of kingdom**" appears in Esther (1:11) it has a prefix letter *beit* (בְּכֶתֶר מַלְכוּת) preceding it. With this additional *beit*, the *gematria* of the phrase (בְּכֶתֶר מַלְכוּת) becomes 1118, the value of our essential statement of faith, "Hear O' Israel, *Havayah* is our God, *Havayah* is one" (Deuteronomy 6:4). The prefix *beit*, whose value is 2, also foreshadows the 2 additional appearances of the phrase in the Book of Esther.

►►$64 = 8^2$. The normative *gematria* of, "**God created**" (בָּרָא אֱלֹהִים) is $289 = 17^2$. The difference between the normative and ordinal values of this same phrase is then $225 = 15^2$. Mathematically, the normative value (289) and the ordinal value (64), and their difference (225) comprise a Pythagorean triplet {8, 15, 17}, since $8^2 + 15^2 = 17^2$.

►►► Together, the value of "**the attribute of judgment**" and "**the attribute of mercy**" (מִדַּת הַדִּין מִדַּת הָרַחֲמִים) is $1260 = $ ◆35 (the diamond, or double triangle of 35). The value of the initial letters (מ ה מ ה) is 90, also a diamond number, specifically: $90 = $ ◆9.

1260 itself is a multiple of 90: $1260 = 90 \cdot 14$, meaning that since the initials equal $1 \cdot 90$, the remaining letters equal $13 \cdot 90$.

1260 can also be restated as $26 \cdot 45$, or *Havayah* times Adam (אָדָם), stressing the embodiment of both God's attributes of judgment and mercy in man.

▶ The World of Creation corresponds to the *sefirah* of understanding, out of which, the *Zohar* (II, 175b) states, **harsh judgments** emit.

numbering. Here too we find the attribute of compassion—originating from the World of Emanation—described as a higher logical foundation for reality than that of judgment. Indeed, when considering nature as based on logical and harsh laws, it is easy to see the affinity between the World of Creation, ordinal numbering and judgment◀ and the subsequent necessity for a quantum leap in sustainability provided by Divine compassion originating in the World of Emanation.

Before completing this point, let us note that the difference between "compassion" (רַחֲמִים), 298 and "judgment" (דִּין) 64, is 234.▼▼ 234 = 9 · 26. These two numbers, 9 and 26 themselves relate to judgment and compassion. 26 is of course the value of God's essential Name, *Havayah*, also known as, "the Name of compassion." Indeed, the 13 Principles of Divine Mercy[4] or compassion begin with a twofold repetition of the Name *Havayah*.

How the number 9 relates to judgment in this case requires some further elaboration. Judgment is considered the building block of the *sefirah* of kingdom, or put in more modern terms: the very foundation of a state is its ability to maintain justice, or law and order, among its populace. Yet, as noted, an entity based on judgment alone, without

---

4. Exodus 34:6-7.

▶▶ After the deluge, God first of all strengthened nature's judgmental aspect by ensuring its rigidity. The Torah (Genesis 8:22) describes this as nature not having a **Sabbath**, meaning that the laws of nature should not take a break, as it were.

The Hebrew word for Sabbath is Shabbat (שַׁבָּת) and its *gematria* is 702. Since it possesses three letters, the average value of each letter is 234, the difference between "compassion" (רַחֲמִים) and "judgment" (דִּין). Thus, where nature is associated with judgment, the Sabbath is associated with compassion. **Adding** the value of "**compassion**" (298) **to the Sabbath** (702) gives 1000, a number most closely associated with the Shabbat as it alludes to the 1000 lights given to Moses at Mt. Sinai (the giving of the Torah was on Shabbat),

subsequently taken from him following the sin of the Golden Calf, and returned to his soul root every Shabbat.

The Sabbath is God's special gift to the Jewish people one of whose unique national traits is compassion. So special is the Sabbath to Jews, that a non-Jew is prohibited from keeping it (even voluntarily). Indeed, in regard to a Jewish state in the Land of Israel, the addition of compassion is achieved by the entire Jewish people keeping the Shabbat.

Adding 234 (transformation of judgment into compassion) to 702 (the value of "Shabbat") gives 936 = 12 · 78. 78 is the value of "bread" (לֶחֶם) and multiplying it by 12 alludes to the *Arizal's* secret of conducting the Shabbat table with 12 loaves of bread.

compassion, is unsustainable. The sages remark that even the Jewish state of the Second Temple period was destroyed, "because only the judgment of the Torah was administered—for they did not add compassion to their judgment."[5] Thus, even though justice and statehood must be founded on judgment, they must actually, in retrospect, be re-initiated upon a foundation of compassion.

The addition of compassion to kingdom depends on the 9 *sefirot* preceding kingdom, with each *sefirah* contributing mercy and compassion from the light of *Havayah*—God's Name of compassion—that radiates within it.▸ Specifically, adding the lights of compassion of the 9 preceding *sefirot* to kingdom is driven by the 9th *sefirah*, foundation. Foundation is described in the central verse describing the *sefirot* as "for all that is in the heavens and the earth." The first two words of this description, "for all" (כִּי כֹל) actually have the same value as "foundation" (יְסוֹד). But, the description further implies that foundation acts so that "all that is in the heavens" also be found in "the earth," where earth refers to the *sefirah* of kingdom. It should now be clear that the product of 9 and 26 alludes to the 9 *sefirot* from crown to foundation radiating lights of compassion, the light of the Name *Havayah*, whose numerical value is 26, resulting in the transformation of kingdom's judgmental attribute into mercy.

Another beautiful allusion to the role that foundation plays in this transformation of kingdom can be seen by focusing on the figure of Joseph, considered the archetypal soul of foundation—i.e., the Biblical persona best representing the qualities of foundation. Joseph's name in Hebrew actually means "to add," another statement of foundation's role of adding all that is in the higher *sefirot* into kingdom.

We can also look at Joseph from a numerical perspective. The *gematria* of "Joseph" (יוֹסֵף) is 156. If we write the filling of his name's letters, we get: יוד ואו סמך פא.▸▸ The letters used to fill the original root letters are וד או מך א, whose sum is 78, one half of 156, or one half

---

5. *Bava Metzi'a* 30b.

▸ The **Divine Names of the *sefirot*** are actually the names describing their vessels. But, the light within is in fact the light of the Name of *Havayah*, the Name of compassion. For more, see *What You Need to Know About Kabbalah*, pp. 139ff.

▸▸ In filling the letters of Joseph, we have used the filling known as **the *mah* filling** (מִלּוּי מה), wherein the letter *hei* (ה) is filled as הא and the letter *vav* (ו) as ואו. The *mah* filling is referred to by this name because the word *mah* (מה) is the value of God's essential Name, *Havayah*, when filled in this manner: יוד הא ואו הא = 45.

45 is also the value of "Adam" (אָדָם), i.e., "man," implying that of the various different filling schemes, the *mah* filling is most symbolic of the human state. Joseph is specifically connoted as simply "man" in Numbers 9:6 (see *Sukah* 25a). Furthermore, the value of "Adam" (אָדָם) in cumulative numbering (see next chapter) א אבגד אגבדהוזחטיכלמ is 156, the value of "Joseph" (יוֹסֵף) in normative numbering.

of Joseph (illustrating a whole and a half relation between Joseph's name and its filling). But, this also means that the full value of Joseph's name when filled is 156 ⊥ 78 = 234, exactly the difference between compassion and judgment. In other words, when we add 234—the value of Joseph's filled name—to the word "judgment" (דִּין), symbolic of Joseph's *sefirah*, foundation, acting in full to augment judgment, the sum comes to 298, the value of the word "compassion" (רַחֲמִים). Stated in more mathematical notation,

$$דִּין \perp יוד ואו סמך פא = רַחֲמִים$$

Pathways and gates

The value of the entire first verse in reduced numbering is 82. The reduced value of the first phrase (as defined by the *etnachta* cantillation mark), consisting of the first 3 words, "In the beginning God created" (בְּרֵאשִׁית בָּרָא אֱ־לֹהִים) is 32, alluding to the 32 Pathways of Wisdom[6]. The reduced value of the second phrase, consisting of the final 4 words, "the heavens and the earth" (אֵת הַשָּׁמַיִם וְאֵת הָאָרֶץ) is 50, alluding to the 50 gates of understanding engraved in the fabric of creation.[7] Realizing this, we are reminded of the beautiful verse, "God founded the earth with wisdom, He set the heavens upon understanding,"[8] describing that creation is based upon both wisdom and understanding together.

We can take this analysis another step further. In the human form, wisdom and understanding correspond to the right and left lobes of the brain. Recall that in chapter 1, we saw that each of these phrases contains 14 letters, thus alluding to a verse from Isaiah describing the heavens and earth as being created by God's figurative right and left hands. The value of the word "hand" (יָד), as we saw is 14, thus both hands allude to 28, the total number of letters in the Torah's

---

6. See *Sefer Yetzirah* 1:1 and its commentaries. See also Rabbi Moses Cordovero, *Sefer Hapardes*, *Sha'ar 12*.
7. In the *Pardes*, *Sha'ar 13*.
8. Proverbs 3:19.

first verse. So here we have another parallel phenomenon with the right and left intellectual faculties (wisdom and understanding▸) being God's tools for creation.

Now, if we calculate the final-reduced value of each half of the first verse, the first half comes to 14 (a self-reference, since 14 is the number of letters in each half) and the second half to 23. Together, the final reduced value is 37, which we saw is the ordinal value of "wisdom," by itself.

Since the reduced values of the verse's two halves are 32 and 50, it is easy to see that the final-reduced value of each half is 5, alluding once again to the 5 fingers on each figurative hand with which God created reality. 5 on the left and 5 on the right also suggests the two Tablets of the Covenant given to Moses at Mt. Sinai. Each tablet contained the text of 5 of the Ten Commandments.

The final-reduced value of the entire first verse of the Torah is 1, the same as the final-reduced value of wisdom. Although the Torah begins the account of creation with the letter *beit* (ב), whose value is 2, the final-reduced value of the first verse reveals that in essence it is the ultimate, absolute "One"—"*Havayah* is one"—that creates all of reality. Reality, which seems to be a plurality, begins with the number 2, the smallest plurality possible, but underlying the apparent plurality lies the hidden truth of God's oneness.

When worlds diverge

Let us now turn to look at the value of the same word in different numbering schemes, i.e., its value in different worlds. Take God's essential Name, *Havayah*. Its value in both normative and ordinal numbering is the same, 26; meaning that it is the same in both the world of Emanation and the world of Creation.

> *A word composed only of the letters from* alef *to* yud
> *has identical normative and ordinal values.*

In the case of *Havayah*, we find an example of the equality between its

▸ As discussed earlier (p. 2), the **28 letters in the Torah's first verse** divide into two groups of 14 letters, where 14 is the value of "hand" (יָד) and alluding to the parallel verse describing creation, "Indeed, My hand has laid the foundation of the earth, and my right hand has spread out the heavens" (Isaiah 48:13). The two figurative hands mentioned in this verse correspond to wisdom and understanding. Moreover, the sum of "wisdom" (חָכְמָה) and "understanding" (בִּינָה) is 140, or 10 times the value of "hand" (יָד).

normative and ordinal values used as a cornerstone for explaining the seeming redundancy of *Havayah* in the 13 Principles of Divine Mercy, which begin with the phrase, "*Havayah, Havayah.*"[9] In the Masoretic text, there is a vertical line (פְּסִיק טַעֲמָא) that separates the two Names *Havayah*. It is explained in Kabbalah that this vertical separator is the secret of the "curtain" (מָסָךְ), symbolizing the contraction (צִמְצוּם) of Divine revelation between Divine Emanation and Creation, themselves symbolized by the two Names *Havayah*.

While the value of *Havayah* in both worlds is the same, the next word that appears in the text, the Name *Kel* (אֵ־ל), the first of the 13 Principles of Mercy, equals 31 in normative *gematria* but 13 in ordinal *gematria*. 13, in this respect, serves as a self-reference to the understanding that there are 13 Principles of Mercy, something that is not at all easily gleaned from the text itself. The fact that the Name *Kel* here is 13 in ordinal *gematria* of course relates to the fact that the second *Havayah* is considered to be in the World of Creation, whose numbering system is ordinal.

So far, we have seen the value of *Havayah* in the two higher Worlds, Emanation and Creation. As stated earlier, the four Worlds correspond to the 4 letters of *Havayah* (יהוה), allowing us to extend the correspondence between the 4 *gematria* systems and the Worlds to the 4 letters of *Havayah*, like so:

letter of *Havayah*	type of *gematria*	values	World
י (*yud*)	normative (הֶכְרֵחִי)	1…10, 10…100, 100…400	Emanation
ה (*hei*)	ordinal (סְדוּרִי)	1…22	Creation
ו (*vav*)	reduced (קָטָן)	1…9 cumulative	Formation
ה (*hei*)	final-reduced (קָטָן אַחֲרוֹן)	1…9	Action

We therefore continue by looking at *Havayah*'s value in reduced and final-reduced numbering, the systems corresponding to the two lower Worlds. The value of *Havayah* in the world of Formation

---

9. Exodus 34:6.

(reduced numbering), is 17 and in the world of Action (final-reduced value), its value is 8. Now we have found that,

World	type of *gematria*	value of *Havayah*
Emanation	normative (הֶכְרֵחִי)	26
Creation	ordinal (סְדוּרִי)	26
Formation	reduced (קָטָן)	17
Action	final-reduced (קָטָן אַחֲרוֹן)	8

The sum of these four values is 77, the normative *gematria* of each of two words used to describe the nature of the Name *Havayah* in the verse, "The Name *Havayah* is a mighty tower"[10] (מִגְדַּל עֹז שֵׁם יהוה). The normative value of both "tower" (מִגְדָּל) and "mighty" (עֹז) is 77! In fact, the value of this entire phrase (מִגְדַּל עֹז שֵׁם יהוה) is 520, a multiple of *Havayah*, 520 = 20 · 26! So is the reduced value of the entire phrase, which is 52 = 2 · 26!▶

Worlds of maturity and youth

The two systems of *gematria* that most frequently appear in the classical texts of Kabbalah are normative *gematria*—the system that corresponds to the world of Emanation, the world of Divine unity—and reduced numbering—the system that corresponds to the world of Formation, the world of apparent plurality. The state of consciousness of Emanation is considered the final stage of mature consciousness, whereas that of Formation is the consciousness of the youth, who as we saw in the previous chapter, is symbolized by the figure of Chanoch (Enoch).

Different people have different conceptions of what it means to be mature. In Judaism in general, and in the inner teachings of the Torah in particular, maturity is a measure of the degree of unity one sees in reality. Since God is one, in truth all that we experience has a single source. But, immaturity can lead us to experience reality as

▶ The reduced value of the phrase, "The Name *Havayah* is a mighty tower" is 52, or 2 times the value of *Havayah*, 26. Multiplying *Havayah* (26) by 2 in the reduced value here alludes to **the two Names Havayah** that precede the 13 Principles of Divine Mercy.

10. Proverbs 18:10.

a plethora of opposing forces, vying with one another for ultimate sovereignty.

Paradoxically, in respect to numerical representation, the opposite seems to be true. The numerical value in Emanation (normative numbering) is either greater or equal to its value in Formation (reduced numbering). The reason for this is, as explained above, that the numerical value of a word is not to be understood quantitatively alone, but primarily qualitatively. The larger the number, the more "light," the more clarity of differentiation exists between one word and another, between different manifestations of the Creator's ultimate unity. Maturity is measured by the ability to resolve between the numerical values of words or letters, but particularly words. From the perspective of Emanation, there is greater distance between the values assigned to different words, and in general, the normative value of a word is greater than its reduced value (hence the notion of "reduced").

Put another way, absolute unity—and God is absolutely one—is illustrated by the paradoxical ability to have infinite manifestations, as explained in the previous chapter. The less the differentiation, the more the word/object appears to be a reality of its own, with many such realities belonging to the same set/number. In the World of Emanation fewer words will have the same value than in the World of Formation. The clarity that comes with greater differentiation allows the word or object's inner soul—its connection to the unity of the Creator (who continually creates it by means of the letters of its name)—to be revealed.

Let us now see an example of this special relationship between Emanation and Formation. God's Name that appears in the Torah's first verse is *Elokim*. In fact, this Name repeats 32 times throughout the account of the first 6 days of creation, providing a major allusion to the 32 pathways of wisdom with which God created the world. It appears another 3 times in the account of the seventh day of rest, the Shabbat. Again, its first appearance is in the first verse, "In the beginning God [*Elokim*] created…"). The *gematria* of *Elokim* (אֱ־לֹהִים) in normative *gematria* is 86, while its value in reduced *gematria* is 14. When added together, $86 + 14 = 100 = 10^2$, a sign that these two numbers do indeed complement each other, forming a perfect square together. Now, the Name *Elokim* appears in the first half of the verse, "In the beginning God created," which has 14 letters. The second half of the verse also has 14 letters (we've discussed this above). The first half of the verse contains the subject—*Elokim*, and the second half contains the object—the heavens and the earth. Yet, even though the subject, *Elokim*, appears in normative numbering in the first half of the verse alone, its reduced value is referred to a second time in the 14 letters of the verse's second half, the half in which the object appears. Moreover, the first half of the verse, the subject half, reduces in final-value to 14, while the second half, the object half, reduces to 23, as we saw above

Emanation, Creation, and Formation together as one

A more complex example of how the numerical values of a word in different worlds interact can be seen in relation to the Name *Havayah*. In Emanation (normative *gematria*) its value is of course 26. In Creation (ordinal *gematria*), it is 26 and in Formation (reduced numbering) it is 17, and their sum is 43. If we now multiply the sum of *Havayah* in Creation and Formation (representing the consciousness of our souls that have descended into mind and emotions), 43, by 26, its value in Emanation (in purely Divine consciousness) the product is 1118! This is a truly amazing phenomenon, as 1118 is the value of our essential statement of faith in God's unity, "Hear O' Israel *Havayah* is our God, *Havayah* is one"[11] (שְׁמַע יִשְׂרָאֵל ייהוה אֱ־לֹהֵינוּ ייהוה אֶחָד)!

Also note that the value of *Havayah* in the world of Action (final-reduced value), 8, does not appear in this relationship, for in the world of Action the Presence of *Havayah* is concealed. This is the symbolic meaning of the word "even" appearing in the verse, "…

---

11. Deuteronomy 6:4.

I have created, I have formed, even have I made."[12] The "even" suggests that the consciousness of the world of Action is severed from that of the three higher worlds (as explained in the previous chapter).

However, the word "even" in this verse from Isaiah can also be interpreted as "even though" (as in the idiom, אַף עַל פִּי, in Hebrew). The verse would then be implying that *even though* the World of Action completely conceals the Presence of Divinity, it is still one of the Four Worlds, and as stated by the Magid of Mezritch, "*Atzilus is auch do,*" meaning that the World of Emanation is also here, where here refers to the World of Action. Regardless of His total concealment, God is present in every aspect of the World of Action.

This encourages us to sum the value of *Havayah* in all three lower Worlds (from Creation to Action) giving us, $8 + 17 + 26 = 51$. Proceeding to multiply this number by *Havayah* itself (26), we get, $51 \cdot 26 = 1326$, a special number both mathematically and in the Torah. Mathematically, $1326 = \triangle 51$. In the Torah, 1326 is the value of the third and culminating verse of the Priestly Blessing, "May *Havayah* bestow His countenance upon you and give you peace" (יִשָּׂא יְהוה פָּנָיו אֵלֶיךָ וְיָשֵׂם לְךָ שָׁלוֹם). The peace mentioned in this verse is all-inclusive peace, extending from the higher Worlds especially down to the World of Action, as we say at the culmination of the *Amidah* prayer, "He who makes peace above [in the World of Emanation] shall make peace upon us [in the three lower Worlds]."

---

12. Isaiah 43:7.

MANKIND'S FIRST FAMILY

One of our observations has been that the ordinal value of "wisdom" (חָכְמָה) is 37, also the normative value of the word "vanity" (הֶבֶל). What more can we say about this relationship?

The Hebrew word for "vanity" (הֶבֶל) is also the proper name of Adam and Eve's second son[1] known as Abel in English. The verses describing the birth of Adam and Eve's first two children read,

> *And Adam knew his wife Eve, she conceived and bore Cain….*
> *She again bore his brother Abel…*

וְהָאָדָם יָדַע אֶת חַוָּה אִשְׁתּוֹ וַתַּהַר וַתֵּלֶד אֶת קַיִן...
וַתֹּסֶף לָלֶדֶת אֶת אָחִיו אֶת הָבֶל...

We may surmise that the numerical relationship between "Abel" and "wisdom" suggests that, of the two children, Abel and Cain, Abel was the wiser. Going a step further, we can add that Abel's superior wisdom, which translates and motivates a sense of self-nullification before the Creator, was the underlying reason for Cain's hatred towards him ultimately leading Cain to murder his younger brother.

In regards to our mathematical analysis of the Torah's first verse, this, our last observation, charts our course into a numerical exploration of mankind's first family (Adam, Eve, and their children) and how they relate back to Genesis.

Eve and wisdom

The first thing we note is that the numerical value of Eve (חַוָּה), the first woman, is 19. This is true in normative, ordinal, and reduced numbering. In final reduced numbering, her name reduces to 1, making the sum of all four levels equal to 58, the value of the word

---

1. Genesis 4:2.

"grace" (חֵן). But, 19 is the midpoint of 37, the value of "Abel" (הֶבֶל), Eve's second son. Abel's name in ordinal *gematria* actually equals 19, the value of his mother's name "Eve" (in normative, ordinal, and reduced numbering). 19 is also the reduced value of "wisdom." Turning to normative *gematria* we can now state that "Eve" (19) is the midpoint of "Abel" (37) who in turn is the midpoint of "wisdom" (73), or in more mathematical notation,

$$\text{חָכְמָה} > \text{הֶבֶל} > \text{חַוָּה}$$

All this reveals the essential relationship between wisdom, Abel, and Eve.

Abel reincarnated

After Abel's murder, Adam and Eve gave birth to a third child, whom Eve called *Shet* (Seth in English). The Torah relates,

> *And Adam knew his wife again*◄ *and she bore a son and she called his name Seth, for God has set me another seed in place of Abel whom Cain slew*[2]

וַיֵּדַע אָדָם עוֹד אֶת אִשְׁתּוֹ וַתֵּלֶד בֵּן וַתִּקְרָא אֶת שְׁמוֹ שֵׁת כִּי שָׁת לִי
אֱ-לֹהִים זֶרַע אַחֵר תַּחַת הֶבֶל כִּי הֲרָגוֹ קָיִן

The Torah explicitly states that it was clear to Eve that Seth had taken the place of her murdered son Abel. In more mystical terminology, Seth is considered Abel's reincarnation.▼▼

---

2. Ibid. 4:25.

► The word "**again**" (עוֹד) equals 80. The product of the three letters that spell "again" is 70 · 6 · 4 = 1680, itself a multiple of 80, specifically, 1680 = 21 · 80, where 21 is the value of "I shall be" (אֶהְיֶה), suggesting that Seth is Abel's essence saying, "I shall be again."

Interestingly, the verse's first word, "knew" (וַיֵּדַע), comprises the 3 letters of "again" (עוֹד) plus an additional *yud* (י), so if we take the product of its letters, we will get 10 times that of "again," namely 16800. Of course, it follows then that 16800 = 210 · 80 (10 times the previous product of 21 · 80). But, 16800 is also the product of 105 and 160, where 160 is both the *gematria* of "Cain" (קַיִן) and the product of the three letters of "Adam" (אָדָם), the word in between "knew" and "again."

▼▼ The Torah relates that idolatry began in the generation of Seth's son, Enosh (Genesis 4:26; see *Rashi* there). With a bit of reflection, the cause and effect between the realization that Seth was a **reincarnation** of Abel and the advent of idolatry becomes clear.

There are two possible views of man: monist or dualist. The dualist stance holds that man is composed of two entities that make up his being; they are usually referred to as matter and spirit, or body and soul. The monist stance (the view held by various ancient philosophers and the core of the materialist-reductionist view adopted by many modern thinkers) argues that man is composed of only one entity, the matter that makes up his body. Even if Adam and Eve had a dualist view of themselves, the notion that Divine spirit could animate, or be present in an inanimate object was not yet warranted. But, when their descendants saw that Abel's spirit had returned to occupy a second body, in Seth, this certainly placed weight on the argument that there exists spirit that could be moved from one body to another, and if this was true of a human body, if the spirit was great enough (like the spirit of the planets, considered much greater than the spirit of man), it might be able to inhabit some inanimate matter. This very leap of thought was what led to idolatry in the generation following Seth.

The power of generations

To understand the special role played by Seth, note that in Hebrew the proper name "Seth" (שֵׁת) and the verb "has set" (שָׁת) in the above verse have identical spelling and meaning. Interestingly, even in English they share a phonetic similarity. The particular "setting" alluded to in Seth's name is the setting, or placement of the *sefirah* of foundation, setting the corner stone of humanity and setting the foundation for all coming generations. The *sefirah* of foundation relates in particular to the "rectification of the covenant" (תִּקּוּן הַבְּרִית), the rectification of the human sexual drive in man, the foundation for a father bearing a son and for each generation bearing the next.

The phrase best describing this in the Bible is, "A generation comes and a generation goes"[3] (דּוֹר הֹלֵךְ וְדוֹר בָּא). The *gematria* of this phrase is 484 = 22^2. 484 is also the *gematria* of the phrase describing Abel's occupation as a shepherd, "Abel became a shepherd"[4] (וַיְהִי הֶבֶל רֹעֵה צֹאן). A shepherd's sheep are considered to embody the souls of his forthcoming progeny.[5] Adding 484, the power of procreation in foundation (and the essence within Abel's shepherding) to the value of "Seth" (שֵׁת), which is 700, we get 1184 = $32 \cdot 37$, where 37 is once more the *gematria* of Abel. Seth indeed is in place of Abel. Moreover, the *gematria* of the words, "set me" (שָׁת לִי) is 740 or $20 \cdot 37$, again a multiple of "Abel."

Abel, Seth, and consummate love of God

Let us place the three letters of the name Abel (הֶבֶל) between the two letters of the name Seth (שֵׁת). This is symbolic of Abel's soul entering Seth's body. What emerges from this unification is the word שַׁלְהֶבֶת (pronounced: *shalhevet*), which means "flame." This first reincarnation of souls symbolizes the ongoing flame

---

3. Ecclesiastes 1:4.
4. Genesis 4:2.
5. *Bereisheet Rabbah* 73:8. See *Shlah, Vayeitzei, Torah Or* 5.

of souls (which are called candles of God[6]) from generation to generation. The value of the word "flame" (שַׁלְהֶבֶת) 737, is equal to the value of the phrase, "[You shall love *Havayah* your God] with all of your heart, and with all your soul, and with all your might" (וְאָהַבְתָּ אֵת ה' אֱ-לֹהֶיךָ] בְּכָל לְבָבְךָ וּבְכָל נַפְשְׁךָ וּבְכָל מְאֹדֶךָ),[7] illustrating numerically that consummate love of God is the soul's eternal flame.

Cumulative numbering

Looking at the four words, "another seed◄ in place of Abel" (זֶרַע אַחֵר תַּחַת הֶבֶל), we see that their value is 1331 = 11³. 1331, a cubic number, is also the *gematria* of "Mashiach" in another central numbering system found in Kabbalah, called *meespar keedmee* (מִסְפָּר קַדְמִי),[8] which we translate as either "primordial numbering" or "cumulative numbering."▼▼

Returning to our phrase, we see that the final letters are ת, ר, ע, and ל, whose sum is 700, the value of "Seth" (שֵׁת)! But, since 700 divides into 4, this means that the average value of these four letters is 175, the lifespan of Abraham, the first Jew.[9] As we shall see, our analysis is leading up to a mathematical comparison between mankind's first nuclear family and Judaism's first nuclear family, begun by Abraham.

---

6. Proverbs 20:27.
7. Deuteronomy 6:5.
8. See *Pardes Rimonim* 30:8b; *Pri Eitz Chaim, Sha'ar hashabat*, 4.
9. Genesis 25:7.

► The final letters of the words "another seed" (זֶרַע אַחֵר) spell **Er** (ער), Judah's firstborn son who sinned and was killed by God. In the words of the Torah (Genesis 38:7), "And Er, Judah's firstborn was evil in the eyes of God" (וַיְהִי עֵר בְּכוֹר יְהוּדָה רַע בְּעֵינֵי הוי'). "Evil" (רע) is Er (ער) spelled backwards.

The filling of the two letters of "Er" (ער) is עין ריש and their value is 640, also the gematria of Tamar (תָּמָר), Er's wife. Thus, we have found that Er's wife represents the manifestation of his full potential. His sin (which is why God found him to be evil) was that he was not willing to impregnate her, thereby realizing *her* potential.

640 is also 4 times 160, the value of Cain (קַיִן). Since in Hebrew, Cain has 3 letters, 4 times the value of Cain is equal to what is known as the "frontside and backside" (פָּנִים וְאָחוֹר) calculation of Cain, which is: ק קי קין קין יו ן. It is easy to convince ourselves that the value of the frontside and backside calculation of a word is always equal to its value times 1 more than the number of letters in the word. Just the frontside of Cain, קין יו ן, equals 270, the value of Er (ער).

►► How does **cumulative numbering** (meespar keedmee) work? In this numbering system, each letter is assigned the cumulative sum of all the letters from *alef* up to and including the letter itself. The values assigned to the letters in cumulative numbering are thus,

letter	name	value	letter	name	value	letter	name	value	letter	name	value
א	alef	1	ז	zayin	28	מ	mem	145	ק	kuf	595
ב	bet	3	ח	chet	36	נ	nun	195	ר	reish	795
ג	gimel	6	ט	tet	45	ס	samech	255	ש	shin	1095
ד	dalet	10	י	yud	55	ע	ayin	325	ת	tav	1495
ה	hei	15	כ	kaf	75	פ	pei	405			
ו	vav	21	ל	lamed	105	צ	tzadik	495			

Indeed, calculating the cumulative value of "Mashiach" (מָשִׁיחַ), we get 145 + 1095 + 55 + 36 = 1331, the same as the value of the phrase, "Another seed in place of Abel" (זֶרַע אַחֵר תַּחַת הֶבֶל).

Another example: the primordial/cumulative value of this phrase's four final letters, ע ר ת ל is 325 + 795 + 1495 + 105 = 2720, or the product of "good" (טוֹב), 17, and Cain (קַיִן), 160, thus alluding to Cain's future rectification, something that will occur with the coming of Mashiach.

▸ When considered together, **Seth and Abraham** yield some noteworthy numerical phenomena. Adding Seth (שֵׁת), 700 to Abraham (אַבְרָהָם), 248 gives 948, making the average value of the two names equal to 474, the *gematria* of "knowledge" (דַעַת), a word that is also translated as "consciousness," mankind's (Adam's) essential attribute.

Seth lived 912 years and Abraham 175. Adding the two names to their life-spans gives us 2035 = 55 · 37, where 55 is the value of "all" (הַכֹּל) and 37, the value of "vanity," or "Abel" (הֶבֶל), alluding once more to the Ecclesiastes' opening statement, "...All is vanity."

Adding the value of Seth, 700 to his life-span, 912 gives 1612 = 26 · 62, making it a multiple of God's essential Name, *Havayah*. Note that 1612 is also the *gematria* of the words, "She conceived and bore Cain" (וַתַּהַר וַתֵּלֶד אֶת קַיִן).

One more point about 1612. It is the *gematria* of the names of the 6 emotive *sefirot* (also called the "six extremes," or קְצָווֹת), "loving-kindness, might, beauty, victory, acknowledgment, and foundation" (חֶסֶד גְּבוּרָה תִּפְאֶרֶת נֶצַח הוֹד יְסוֹד). As noted, Seth corresponds to foundation, and therefore this last equality reflects that at times, foundation is considered to include all six emotive *sefirot*.

Abraham is truly a messianic figure, a soul sent to the world to redeem mankind, just as Seth was seen by his mother Eve as the possible redeemer (and rectifier) of a generation▴ in which a man could murder his own brother.

Adam's first rectified act of (pro)creation

Taking stock of the full verse describing Seth's birth, we see that in the original Hebrew it contains 22 words and 66 letters.

וַיֵּדַע אָדָם עוֹד אֶת אִשְׁתּוֹ וַתֵּלֶד בֵּן וַתִּקְרָא אֶת שְׁמוֹ שֵׁת כִּי שָׁת לִי
אֱ־לֹהִים זֶרַע אַחֵר תַּחַת הֶבֶל כִּי הֲרָגוֹ קָיִן

Both 22 and 66 divide by 11 and 66 = △11. Furthermore, the complete verse divides into two parts, each with 11 words. The ratio of words to letters▾▾ is 1:3, a very rare ratio in the Torah.▸▸▸

▸▸▸ In fact, there are only 26, the value of God's essential Name, *Havayah*, verses that demonstrate this ratio of words to letters.

▸▸ Let's look at the verses with **the lowest and highest word-to-letter ratios**. The lowest ratio in the entire Pentateuch is Deuteronomy 14:18, וְהַחֲסִידָה וְהָאֲנָפָה לְמִינָהּ וְהַדּוּכִיפַת וְהָעֲטַלֵּף. With 5 words and 32 letters, each word on average contains 6.4 letters. The verse with the highest word-to-letter ratio is Exodus 21:24, עַיִן תַּחַת עַיִן שֵׁן תַּחַת שֵׁן יָד תַּחַת יָד רֶגֶל תַּחַת רָגֶל. It has 12 words and again 32 letters; every word is, on average, merely 2 and 2/3 letters long.

Note that the verse with the highest word to letter ratio and the verse with the lowest both have 32 letters, where 32 alludes to the word "heart" (לֵב), suggesting two types of comprehension in the heart. Simply put, words represent an inclusive principle that contains particular details—the letters. Thus, the high word-to-letter ratio of Deuteronomy 14:18 illustrates an "understanding heart" (לֵב נָבוֹן), since the *sefirah* of understanding fosters comprehension based on many details (see Proverbs 16:14 and 18:15 for more on this phrase). The low word-to-letter ratio of Exodus 21:24 illustrates a "wise heart" (לֵב חָכָם), since the *sefirah* of wisdom is associated with comprehension based on general principles. Both understanding and wisdom are intrinsically connected to the heart in Kabbalah. Wisdom contains 32 pathways, the value of "heart." Understanding is described (introduction to the *Tikunei Zohar*) as the *sefirah* contained in the heart (בִּינָה לִיבָּא).

The value of the two verses together is 5797, and thus the average value of each of the 17 words is 341, the value of "the shield of Abraham" (מָגֵן אַבְרָהָם). The first and last letter in each word equal together 5177 (the value of *Kel* times "*Havayah* is our God,

▷▷▷

► The first two words of the verse, "And **Adam knew**" (וַיֵּדַע אָדָם), equal 90 and 45 respectively, a ratio of 2:1—"a whole and a half"—the relation and union of husband and wife in the rectified act of procreation. When these two words are written in full (וו יוד דלת עין אלף דלת מם) they equal 1221, the product of 33 and 37, the value of Abel (הֶבֶל), teaching us that Adam and Eve's second act of procreation serves to rectify Abel's soul!

(Note also that the letters themselves divide nicely into two groups, each a multiple of 37. The first, middle, and last letters, whose filling is וו עין מם equal 222, or 6 times 37. The filling of the four remaining letters יוד דלת אלף דלת equals 999, or 27 times 37.)

The first 7 words in the verse form a complete phrase that contains 22 letters (a reference to the total 22 words of the entire verse), "Adam knew his wife again and she bore a son" (וַיֵּדַע אָדָם עוֹד אֶת אִשְׁתּוֹ וַתֵּלֶד בֵּן). Its *gematria*, is 1815, the total number of letters in the complete first account of creation, i.e., including all seven days of creation. The first account of creation ends with the words, "God created to do" (בָּרָא אֱ-לֹהִים לַעֲשׂוֹת), suggesting that it is mankind's task to emulate God's act of creation by procreating too. "And Adam knew his wife again and she bore a son" was mankind's first rectified act of procreation and thus was the first human act to reflect God's creation of the world in seven days.

Expanding our focus to the entire first half of the verse, "And Adam knew his wife again and she bore a son and she called his name Seth" (וַיֵּדַע אָדָם עוֹד אֶת אִשְׁתּוֹ וַתֵּלֶד בֵּן וַתִּקְרָא אֶת שְׁמוֹ שֵׁת). Its value is $3969 = 63^2$, where 63 is 3 times the value of God's Name *Ekyeh* (אֶהְיֶה), as will be explained below. The *gematria* of the next eight words in the verse, "for God has set me another seed in place of Abel" (כִּי שָׁת לִי אֱ-לֹהִים זֶרַע אַחֵר תַּחַת הֶבֶל) is $2187 = 3^7$.

The final word of the verse is Cain (קָיִן), whose value is 160. The value of Cain's name, 160, is also the product of the three letters of his father's name, Adam (אָדָם), or 1 (א) times 4 (ד) times 40 (ם). 160 is also a divisor of the complete verse's value, which is $6560 = 41 \cdot 160$, where 41 is the value of "mother" (אֵם), thus, the entire verse is a multiple of the two words "Cain" and "mother," which in Hebrew can be read as "mother of Cain," another allusion to Eve.

Since the entire verse is 41 times "Cain," it follows that excluding

▷▷▷ *Havayah* is one"), which means that the middle letters equal 620, the value of "crown" (כֶּתֶר), suggesting that the two verses are the supernal crown and the lower crown of the Torah, in terms of word-to-letter ratio.

Since together the two verses have 64 letters, we can write them in the form of a square of 8:

```
ע י נ ת ח ת ע י
ן ש ו ת ח ש ת ו
י ד ת ח ת י ד ר
ג ל ת ח ת ח ר ג ל
ו ה ח ח ס י ד ה ו
ה א נ פ ה ל מ י
נ ה ו ה ה ד ו כ י
פ ת ו ה ע ט ל ף
```

The letters on the two diagonals of this square equal 1833, the value of 3 times "Torah" (תּוֹרָה).

the final word, "Cain," the *gematria* of the other 21 words in the verse is 6400 = 40 · 160. But, 6400 is a perfect square: 6400 = 80², where 80 is the *gematria* of the third word of the verse, "again" (עוֹד), which we will return to look at again below.

Another perfect square can be found in the four words, "his wife and she bore a son" (אֶת אִשְׁתּוֹ וַתֵּלֶד בֵּן), whose value is 1600 = 40² = 10 · 160, or "Cain" (קַיִן). So, just these four words equal one-fourth the value of the first 21 words of the verse.

The first family

From the many striking numerical phenomena found in this verse, we may conclude that all the members of the first family of mankind are most integrally related to one another, or inter-included in one another. As we have seen, their soul-roots continue to live on forever. This motivates us to look at the combined value of the names of the first nuclear human family. We find that adding "Adam" (אָדָם), 45 plus "Eve" (חַוָּה),▶ 19 plus "Cain" (קַיִן), 160 plus "Abel" (הֶבֶל), 37 plus "Seth" (שֵׁת), 700 gives us 961 = 31², a perfect square, again showing the inherent connection between these names.

It is truly remarkable that 961 is also the sum of the names of the first Jewish family as well. Abraham (אַבְרָהָם), 248 plus Sarah (שָׂרָה), 505 plus Isaac (יִצְחָק), 208 = 961! What can we learn from this surprising equality?

Front and back sides of 31

To answer this question, let us go back to the square root of 961, this special number that figures so prominently in both mankind's first family and the first Jewish family. Its square root is 31.

The two simplest words whose value is 31 are either "*Kel*" (אֵל), one of God's Names, or "no" (לֹא). Since the difference between the two words is simply a reversal of their letters, "no" is considered the reverse (אָחוֹר) of *Kel*. This is especially prominent in the *Albam*

▶ Adding the value of **Adam and Eve**, i.e., 45 and 19 gives us 64, another perfect square. 64 is also the value of "God created" (בָּרָא אֱלֹהִים) in ordinal *gematria*, as we saw above, linking Adam and Eve with the very beginning of creation. When we add the ordinal values of "Adam" and "Eve," 18 and 19, we get 37, the value of their son "Abel" in normative *gematria*!

The Torah recounts (Genesis 3:20) that Adam's wife was named Eve because, "She was the mother of all life." The average value of 19, the *gematria* of "Eve" (חַוָּה), and 109, the *gematria* of "mother of all life" (אֵם כָּל חַי) is also 64, the combined value of Adam (אָדָם), 45, and "Eve", 19, suggesting that her bond with Adam is captured in her name.

transformation created by dividing the 22 letters of the Hebrew alphabet into two sets of 11 letters—*alef* through *kaf* in the first set and *lamed* through *tav* in the second—and then pairing the letters from both sets in order:

א	*alef*	⇔	ל	*lamed*		ז	*zayin*	⇔	צ	*tzadik*
ב	*beit*	⇔	מ	*mem*		ח	*chet*	⇔	ק	*kuf*
ג	*gimel*	⇔	נ	*nun*		ט	*tet*	⇔	ר	*reish*
ד	*dalet*	⇔	ס	*samech*		י	*yud*	⇔	ש	*shin*
ה	*hei*	⇔	ע	*ayin*		כ	*kaf*	⇔	ת	*tav*
ו	*vav*	⇔	פ	*pei*						

Thus, the first row in the *Albam* transformation pairs the two letters *alef* and *lamed*. Of course, we could write this pair as either *alef-lamed* or *lamed-alef*. However, there is a difference in nomenclature. When the letter that comes first in the alphabet is the first letter in the transformation, that order is referred to as the front pairing. However, if the letter that comes later in the alphabet appears first in the transformation, that order is called the back pairing. In our case, the *alef* preceding the *lamed*—as in the case of *Kel* (אֵ-ל)—is referred to as the front (פְּנִים) pairing. When the *lamed* precedes the *alef*—as in "no" (לֹא)—that is the reverse or back (אָחוֹר) pairing.

Negation

In the Torah, we find that the first human is associated with the word "no." The description of Eve's creation in order to form the first family with Adam begins with the word "no," "It is not good for man to be alone, I will make for him a help-mate"[10] (לֹא טוֹב הֱיוֹת הָאָדָם לְבַדּוֹ אֶעֱשֶׂה לּוֹ עֵזֶר כְּנֶגְדּוֹ). Moreover, there are 31 (לא) letters in these words spoken by God.◄ These are the only words in the entire account of creation that God says to Himself!

Taking a closer look at these words yields many more findings linking it with the first human family.

• The first and last words in the phrase are "no" (לֹא) and "mate"

───────────

10. Ibid. 2:18.

▶ According to the *Zohar* (III, 12a) this verse is the tenth of **the Ten Sayings**, or Utterances spoken by God to create the world. Being that it is the tenth it corresponds to the tenth *sefirah*, kingdom, also known as the feminine principle (*nukva*).

(כְּנֶגְדּוֹ). Their sum is 114, which is also 6 times 19, the value of "Eve" (חַוָּה). As with any three-letter word, 6 times the value of the word equals the sum of the word's six possible permutations.

- The first, middle, and final words of the phrase describing the motivation for Eve's creation—"It is not good for man to be alone, I will make for him a help-mate"—are: "no" (לֹא), "alone" (לְבַדּוֹ), and "mate" (כְּנֶגְדּוֹ). Their values are 31, 42, and 83 respectively.▶ Their sum thus comes to 156 = 3 · 52, where 52 is the value of "son" (בֵּן), making it also the average value of these three words.▶▶

- The entire phrase equals 1333 = 43 · 31, and since it possesses 31, i.e., "no" (לֹא) letters, this means that the average value of each letter is 43, the sum of two of God's Names, *Havayah* (יהוה), 26, and *Akvah* (אהוה), 17. 17 is the reduced value of *Havayah* as we saw earlier. It is also the value of "good" (טוֹב), the motivation for creating Eve in the first place, for without her, "It is not good...," and with a wife, "one who has found a wife has found good."[11]

The full verse containing the phrase we have been discussing reads, "*Havayah Elokim* said, 'It is not good for man to be alone, I will make for him a help-mate'" (וַיֹּאמֶר יְ־הוָה אֱ־לֹהִים לֹא טוֹב הֱיוֹת הָאָדָם לְבַדּוֹ אֶעֱשֶׂה לּוֹ עֵזֶר כְּנֶגְדּוֹ). It too contains allusions to mankind's first family:

- The *gematria* of the entire verse is 1702 = 2 · 23 · 37, where 37 is once again the value of "Abel" (הֶבֶל).
- The full verse possesses 45 letters, the *gematria* of "Adam" (אָדָם).

What we have gained so far is a deeper understanding of why we associate the square root, 31, of the first human nuclear family with the word "no." The building of the first human family is based on the negation of "aloneness," or even "oneness." Not having found a mate among the creatures, Adam could not bear his solitary being, necessitating his splitting into two halves, a male and a female. But, the negation of solitude lies at the heart of the first human family. The

---

11. Proverbs 18:22.

▶ Note that 42, the value of "alone" (לְבַדּוֹ) is the midpoint of 83, the value of "mate" (כְּנֶגְדּוֹ), suggesting that the ability to find and nurture one's relationship with one's mate hinges on the ability to find one's **essential point of contact** with the Almighty who is described as, "He alone is Himself" (הוּא לְבַדּוֹ הוּא).

▶▶ The value of the phrase describing the motivation for Eve's creation, "It is not good for man to be alone, I will make for him a help-mate" (לֹא טוֹב הֱיוֹת הָאָדָם לְבַדּוֹ אֶעֱשֶׂה לּוֹ עֵזֶר כְּנֶגְדּוֹ) is 1333. The phrase itself contains 31 letters, and 1333 = 43 · 31.

All three numbers appearing in this product, 31, 43, and 1333 are **covenant numbers**—figurate numbers generated by the function,

$$f[n] = n^2 + n + 1.$$

When drawn, they take on the form of 2 triangles drawn apex to apex with a single point in between them. For this reason, we use the notation, $\mathbf{I}n = 2\triangle n + 1$. Here are the first few covenant numbers, starting with n = 0:

In the case of these three numbers, we have:

$$31 = \mathbf{I}5 = 2\triangle5 + 1$$
$$43 = \mathbf{I}6 = 2\triangle6 + 1$$
$$1333 = \mathbf{I}36 = 2\triangle36 + 1$$

desire to escape solitude is perhaps the lowest motivator for a man and a woman to create a marital bond. On the other hand, the desire to serve the Almighty together provides the highest motivation for marital union,[12] as we will see regarding the first Jewish family.

In the same vein, the bearing of children can be an escape from one's mortality. Following the literal reading of the verses, all three of Adam and Eve's children were born after they sinned by eating from the Tree of Knowledge and their consequent punishment of mortality (originally, man was destined to live eternally). It therefore certainly fits that their pursuit of a family was already colored by a misplaced pursuit of immortality vicariously through their children. Thus, marriage and family in Adam and Eve's context bears the distinct mark of negation; first of solitude and then of death.

Affirmation

The first Jewish family, in contradistinction to Adam and Eve, marks a reaffirmation of Divine principles regarding the sanctity and purpose of life. Abraham and his wife Sarah were wed under very different circumstances than Adam and Eve. Sarah was actually Abraham's niece, the daughter of his deceased brother Haran, killed in an attempt to affirm Abraham's belief in monotheism. Both Abraham and his remaining brother Nachor acted with true loving-kindness and took Haran's two daughters as wives. Abraham went further and adopted Haran's only son, Lot.

In respect to children, Abraham and Sarah sought offspring in order to continue Abraham's path of Divine service. Their years of infertility caused them both tremendous distress to the point that Sarah decided to give Hagar, her maidservant to Abraham, hoping that a child begat by Hagar could be raised by Sarah as her own. But, when she realized that Ishma'el, the child born to Hagar was spiritually flawed, she gave up on her dream and forced Abraham to disown him and his mother. Finally, after 14 additional years, Sarah

12. See our volume, *The Mystery of Marriage.*

▶ The *gematria* of the three words, "*Havayah*, the eternal God [*Kel*]" (יהוה א־ל עולם) is 203, the same value as the word "created" (בָּרָא), the Torah's second word. The three letters of "created" (בָּרָא) are a permutation of the first three letters of "Abraham" (אַבְרָהָם).

Moreover, the value of the entire phrase, "And he [Abraham] called there in the Name of *Havayah*, the eternal God [*Kel*]" (Ibid) (וַיִּקְרָא שָׁם בְּשֵׁם יהוה א־ל עולם) is 1202, the same value as the Torah's first three words, "In the beginning God created" (בְּרֵאשִׁית בָּרָא אֱלֹהִים). [1202 is

also equal to the first three letters of "Abraham" (אַבְרָהָם) when the *alef* (א) is set equal to 1000.]

Both equalities allude to the sages' explanation that the world was created through **the conduit of Abraham's soul-root**, a teaching they learn from the word, "[These are the chronicles of the heavens and the earth] when they were created" (Ibid. 2:4) (בְּהִבָּרְאָם), which permutes to spell "with Abraham" (בְּאַבְרָהָם).

---

gave birth to Isaac. Yet, in spite of how important it was for Abraham to have offspring in order to continue his path of Divine service, when commanded by God to sacrifice Isaac, he did not hesitate, demonstrating that his own immortality was not a motivator for having a family.

Clearly, Abraham and Sarah's family was based on loving-kindness connoted by God's Name *Kel*, the name associated with loving-kindness, as we find in the verse,[13] "The loving-kindness of God [*Kel*] is all day long." Indeed, *Kel* is the Name of God Abraham used to publicize monotheism, as the Torah states, "And he [Abraham] called there in the Name of *Havayah*, the eternal God [*Kel*]"[14] (וַיִּקְרָא שָׁם בְּשֵׁם יהוה א־ל עולם). ▲

The content of these two phrases connects with a third phrase, "The world is constructed from loving-kindness"[15] (עולם חֶסֶד יִבָּנֶה). Most significantly, when we count the letters in these three phrases—"The loving-kindness of *Kel* is all day long" (חֶסֶד א־ל כָּל הַיּוֹם), "*Havayah*, the eternal *Kel*" (יהוה א־ל עולם), and "The world is constructed from loving-kindness" (עולם חֶסֶד יִבָּנֶה) ▶▶—we find that together they possess exactly 31 letters, the value of God's Name *Kel* (א־ל), the Name associated with loving-kindness. This provides a powerful allusion to the connection that these three phrases have with Abraham and his family's pursuit of loving-kindness.

In fact, Abraham's (אַבְרָהָם) essence is so full of loving-kindness that we find that his name, whose value is 248, is a multiple of 31:

▶▶ The values of these three phrases "The loving-kindness of *Kel* is all day long" (חֶסֶד א־ל כָּל הַיּוֹם), "*Havayah*, the eternal *Kel*" (יהוה א־ל עולם), and "The world is constructed from loving-kindness" (עולם חֶסֶד יִבָּנֶה) are 214, 203, and 285, respectively. Their sum is $702 = 27 \cdot 26$, which is also the diamond form of 26, which we denote as ◆26.

702 is **the value of "Shabbat"** (שַׁבָּת) and 26 is of course the value of *Havayah*. Note that in all three phrases, the Name *Havayah* appears only once, but this also means that the remaining words equal 676, or *Havayah* (26) squared!

---

13. Psalms 52:3.
14. Genesis 21:33.
15. Psalms 89:3.

▶ The sum of the middle letters in **Abraham, Sarah, and Isaac** (אַבְרָהָם שָׂרָה יִצְחָק) is 505, the value of "Sarah" (שָׂרָה). It follows therefore that the remaining letters equal 456, the *gematria* of "Abraham" (אַבְרָהָם) and "Isaac" (יִצְחָק), implying that Sarah is the inner power uniting the two.

We see this relationship demonstrated again when Abraham and Isaac are on their way to sacrifice Isaac. On their journey to Mt. Moriah, father and son experience the greatest bond between them, something the Torah describes as "the two of them walked together" (וַיֵּלְכוּ שְׁנֵיהֶם יַחְדָּו). The value of these three words is exactly 505, the value of "Sarah" (שָׂרָה).

$248 = 8 \cdot 31$, an equality expounded upon in length in the writings of the Arizal. This means then that the value of "Sarah" (שָׂרָה), 505, and "Isaac" (יִצְחָק), 208 together is also a multiple of 31.◀ Specifically, $505 + 208 = 713 = 23 \cdot 31$.

In summary of this part of our analysis, we have seen why even though the value of the first human family (Adam, Eve, Cain, Abel, and Seth)▼▼ and the first Jewish family (Abraham, Sarah, and Isaac) are both 961, or 31^2, Adam's family alludes to the word "no" (לֹא), while Abraham's family alludes to God's Name, *Kel* (אֵל).

Creation and Seth's birth

Perhaps the most amazing numerical phenomenon regarding the verse describing Seth's birth is seen when we add it to the Torah's first verse, our focus in this volume. Recall that the Torah's first verse, "In the beginning God created the heavens and the earth" (בְּרֵאשִׁית בָּרָא אֱלֹהִים אֵת הַשָּׁמַיִם וְאֵת הָאָרֶץ), equals 2701. The *gematria* of the verse describing Seth's birth is 6560 and together,

---

▶▶Let us make **a *sefirot*-based model of the members of mankind's first family**. The five members of mankind's first family correspond to a basic sub-structure of the ten *sefirot*, as follows:

understanding (בִּינָה)		wisdom (חָכְמָה)
Eve		Adam
	knowledge (דַּעַת)	
coronet of might		coronet of loving-kindness
(עֲטָרָא דִּגְבוּרוֹת)		(עֲטָרָא דַּחֲסָדִים)
Cain		Abel
	foundation	
	Seth	

Adam corresponds to wisdom (חָכְמָה), also called the father principle. Eve, described as "the mother of all life" (Genesis 3:20) corresponds to understanding (בִּינָה), the mother principle. Like parents, the two intellectual powers of wisdom and understanding are considered inseparable.

Knowledge (דַּעַת), the third intellectual faculty divides into two aspects, referred to as two "coronets" (עֲטָרִין): the coronet of loving-kindness and the coronet of might. These two components of knowledge correspond to the two twins Cain and Abel. The coronet of might (the left aspect of knowledge) corresponds to Cain and the coronet of loving-kindness (the right aspect of knowledge) to Abel.

Finally, Seth, whose name as explained in the text means the same as "foundation" in Hebrew, corresponds to the *sefirah* of foundation—the power of knowledge to connect opposites as present in the emotions of the heart.

Both the first union of Adam and Eve to bear Cain and Abel and their second union to bear Seth are referred to by the Torah as acts of knowledge (Genesis 4:1 and v. 25). These two acts of knowledge, as they are described, correspond to the knowing of the mind and the knowing of the heart, respectively. The second union is described more exactly as "[knew...] again" (וַיֵּדַע...עוֹד). Indeed, the value of the word "again" (עוֹד) is 80, the same as the value of "foundation" (יְסוֹד).

$$2701 + 6560 = 9261 \blacktriangleright = 21^3$$

The cube of 21 clearly refers to God's Name *Ekyeh* (אֶהְיֶה), whose value is 21 and which appears exactly three times in the entire Bible, all three in the same verse[16] describing the moment in which God revealed this Name to Moses as the Name of His power to redeem the Jewish people from their bondage in Egypt. More generally, *Ekyeh* is associated with the Kabbalistic mother principle, the power to give birth, which in the context of the Exodus, alludes to the birth of the people of Israel from the Egyptian womb, so to speak.

In passing, we should mention that 21^2, which equals 441, is also associated with the redemption from Egypt. At the beginning of the Torah section of *Va'eira*, God tells Moses that though He had vowed to the Patriarchs that they would inherit the land of Canaan, they had not seen the fulfillment of this oath in their lifetimes. God describes this as, "I did not reveal myself to them with my Name, *Havayah*." *Rashi*, in his commentary on this verse, writes that God is saying that He had not revealed His measure of "truth" (אֱמֶת)—i.e., His power to fulfill His oath—to the Patriarchs, but is now about to reveal it to their offspring, by redeeming them from Egypt and bringing them to the Land of Canaan. The value of "truth" (אֱמֶת), God's power of fulfillment is $441 = 21^2$.

Now, focusing on the number of letters in each verse—the Torah's first verse and Seth's birth—we see that both have a triangular number of letters. The former contains $28 = \triangle 7$ letters, the latter $66 = \triangle 11$. Let us picture what these two verses look like when drawn in the form of two triangles one over the other,

▶ Adding the three verses that make up the Priestly Blessing (Numbers 6:24-6) to this number gives: $9261 + 2718 = 11979 = 11 \cdot 33^2 = 3^2 \cdot 1331 = 3^2 \cdot 11^3$, where we have already seen that **1331** is the value of the phrase, "another seed in place of Abel" (זֶרַע אַחֵר תַּחַת הֶבֶל)

16. Exodus 3:14.

```
                        ב
                      ר   א
                    ת   י   ש
                  א   ר   ב   א
                ל   ה   י   מ   א
              ת   ה   ש   מ   י   ם
            ץ   ר   א   ה   ת   א   ו
      ו   י   י   ד   ע   ד   מ   א   ד   ע   ו   י   א
          ב   ד   ל   ת   ו   ו   ת   ש   א   ת
          ש   ת   א   ר   ק   ת   ו   ו
            ת   י   ש   כ   ת   ש   ו   מ
              ם   י   ה   ל   א   י   ל
                ר   ח   א   ע   ר   ז
                  ב   ה   ת   ח   ת
                    ה   י   כ   ל
                      ו   ג   ר
                        ק
                        ו
```

Looking at this figure, we note that the first and last letters in it spell the word "son" (בֵּן).

The top and bottom 3 letters of this figure are the words "created" (בָּרָא) and "Cain" (קַיִן), which together equal 363 = 3 · 11². But, note that the three top letters also mean "son" in Aramaic.

The first and last words from these two verses, "In the beginning" (בְּרֵאשִׁית) and "Cain" (קַיִן) occupy the top 3 and bottom 2 rows. Their combined value is 1073 = 29 · 37, where 37 is again the value of "Abel" (הֶבֶל)!

Focusing on the center axis of this figure, its letters are, בייהדרלתגן, which equal 714, or 7 times the value of "faith" (אֱמוּנָה). Thus, subtracting their value, 714 from 9261 (the value of the entire figure) leaves 8547, or 7 times 1221, or 33 times Abel (הֶבֶל). Therfore, the entire figure can be understood as suggesting the faith instilled by Abel.

Figuratively, the top row of the triangle of 11 has 4 more letters than the bottom row of the triangle of 7. These 4 additional letters, to the left and the right are וידא. Their value is 21, the cubic root of the value of the entire figure! Indeed, the top row of the triangle of 11 equals 216, itself a cube number, 6^3.

One final analysis of this figure that we would like to do is taking every other line. The first set of lines is:

```
                          ב
                      א   ר
                  ש   י   ת
              ב   א   ר   א
          ל   ה   י   מ   א
      ת   ה   ש   מ   י   ם
  ו   א   ה   ת   א   ר   ע
א   ד   ו   ע   מ   ד   א   ע   ד   י   ו
  ב   ד   ל   ת   ו   ו   ת   ש   א   ת
    ש   ת   א   א   ר   ק   ת   ו   ו   ן
        ת   ש   י   ת   ש   ו   מ
          ם   י   ה   ל   א   י   ל
              ר   ח   א   ע   ר   ז
                  ב   ה   ת   ח   ת
                      ה   י   כ   ל
                          ו   ג   ר
                              י   ק
                                  ו
```

Their value is 5187 or 273 times 19, the value of "Eve" (חַוָּה). The other set of alternating lines in the figure thus equals 4074, or 42 (the Name of God associated with creation) times 97, the value of Meheitavel (מְהֵיטַבְאֵל),[17] the woman considered to be one of the major rectifications of Eve in the Bible, as explained in the writings of the Arizal.

---

17. Genesis 36:39.

THREE AND FOUR

Many speculate as to what number might be at the very heart of creation. What number is it that the universe relies upon the most to function? Some scientists might offer pi—a transcendental number that cannot be fully expressed—as a candidate. Cosmologists might point to the inverse of the fine structure constant, 137, or to some other important physical constant.

So far, we have seen how the first verse of Genesis directly refers to various numbers, most of them quite large (28, 73, 913, 2701, etc.). In this chapter we will see how at a somewhat deeper level, creation depends on the unification of the two numbers 3 and 4.

The first allusion to this interesting suggestion is that the numerical value of the word for "three" (שְׁלֹשָׁה) is 635 and that for "four" (אַרְבָּעָה) is 278 (this type of calculation is known as *meespar meesparee*, in Kabbalah). Their sum is therefore 913, the exact value of the Torah's first word, "In the beginning" (בְּרֵאשִׁית). Thus, the beginning of creation (in particular, the creation of time,▶ "the beginning") depends upon the union of three and four.

Archetypes of 3 and 4

Before continuing our exploration of the connection between 3, 4, and creation, let us look at the symbolism of these two numbers in Torah. Certainly, the most basic symbolic connection between these two numbers is found in respect to the Patriarchs and Matriarchs; the Patriarchs are three—Abraham, Isaac, and Jacob—and the Matriarchs are four—Sarah, Rebeccah, Leah, and Rachel. These two sets of archetypal figures are the most basic and initial association made by a Jew with the numbers 3 and 4.

In the Torah, there is another clear example of 3 and 4 coming together not as two separate sets of souls but as a 3 by 4 matrix (3

▶ Because **the Hebrew calendar** takes into account both the lunar and solar years, mediating between the two based on calculations alone is impossible (because the earth, the sun, and the moon, together form what is known as a 3-body system in physics, ensuring that their motion relative to one another cannot be exhaustively calculated).

Among the guidelines for maintaining a correct lunar-solar calendar is that every cycle of 19 years must include 12 simple years (years with 12 months) and 7 pregnant years (years with an additional 13th month). This guideline is known as the "secret of pregnancy." The value of "pregnancy" (עִבּוּר) is itself "four" (אַרְבָּעָה) [as well as "the hidden light" (אוֹר הַגָּנוּז). The value of "the secret of pregnancy" (סוֹד הָעִבּוּר) is 353, also the value of the words (Psalms 25:14), "The secret of *Havayah* is to those who fear Him" (סוֹד יְיהוה לִירֵאָיו).]

But note that these two numbers 12 and 7 are actually the product and sum of 3 and 4 ($3 \cdot 4 = 12$ and $3 + 4 = 7$). Since the length of the pregnancy cycle, which is 19, is the value of "Eve" (חַוָּה), "the mother of all life" (Genesis 3:20), the secret of the Hebrew calendar is how all of reality becomes pregnant and gives birth 7 times every 19 years.

Another beautiful allusion to the relationship between creation and the secret of the Hebrew calendar can be seen in the value of the idiom, "Hebrew date" (תַּאֲרִיךְ עִבְרִי), which is 913, also the value of the first word of creation, "In the beginning" (בְּרֵאשִׁית)!

columns by 4 rows). This is the layout used to set the stones in the High Priest's breastplate. The Torah writes, "And you shall set in it settings of stones, four rows of stones…"[1] with each row containing three precious stones. On the 12 precious stones were engraved the names of the 12 tribes of Israel.

		אברהם יצחק יעקב
לוי	שמעון	ראובן
זבולן	יששכר	יהודה
גד	נפתלי	דן
בנימין שבטי ישרון	יוסף	אשר

Because in the High Priest's breastplate, 3 and 4 serve to create a two-dimensional object, they carry our comprehension of these two numbers to a deeper level. 3 enumerates the number of columns across, i.e., horizontally across the breastplate, and thus we conclude that the number 3 refers to a relatively horizontal state of consciousness. Since the number 4 enumerates the number of rows down, it refers to a relatively vertical state of consciousness.

Extending this last observation regarding consciousness we can now understand how these two numbers lie at the base of our entire spiritual service of God. Horizontally inclined spiritual consciousness suggests a *forward and inward* movement, particularly in regard to the human psyche and soul. Thus, the number 3 relates to the three *stages* of psychological rectification called: "submission, separation, and sweetening"—the basis of all Divine service as taught by the Ba'al Shem Tov.[2]

Vertically inclined spiritual consciousness is related to the mind and heart's yearning to ascend the hierarchy of Worlds and to

1. Exodus 28:17.
2. *Keter Shem Tov*, 28. See in length in our volume *Transforming Darkness into Light*.

penetrate the deepest mystery of God's essential Name, *Havayah*. Vertical consciousness is thus associated with an upwards and outwards vector of motion. Both the Worlds and God's essential Name, *Havayah* have 4 elements (4 Worlds: Emanation, Creation, Formation, and Action and 4 letters of *Havayah*: *yud, hei, vav, hei*). In summary, we may say that the number 3 is associated with process and the number 4 with hierarchy. ▸

In fact, 3 as process and 4 as hierarchy appear quite clearly in the Torah's first verse. The verse itself contains 7 words that divide grammatically into 3 and 4. The first 3 words, "In the beginning God created" clearly illustrate process (the process of creation), while the final 4 words, "the heavens and the earth" clearly indicate hierarchy (from the heavens to the earth)!

Sefer Yetzirah (the Book of Formation) includes rigorous definitions of a number of different Torah models that are based on 12 elements: the 12 months of the year, the 12 signs of the zodiac, the 12 simple letters of the Hebrew alphabet, etc. Central to our own discussion here of the connection between 3 and 4 is that in each model of 12, the elements are divided into 4 groups of 3. A complete treatment of these models and their 4 by 3 divisions is well beyond our scope here. However, it is important to mention that *Sefer Yetzirah*, as its name implies, also deals with the process of the formation of the worlds, thus suggesting a subtle connection between the first word used by the revealed text of the Torah to describe the process of creation, "In the beginning," and this seminal work of the Torah's concealed dimension.

Still, even in the breastplate, in the Torah's description only the number 4 is mentioned explicitly, whereas the number 3 remains implicit. A more explicit example of 3 and 4 can only be found in the Book of Proverbs, the topic of our next section.

Solomon's 3 and 4

Now that we have looked at some of the most basic symbolic associations of the numbers 3 and 4, we can turn to what is perhaps

▸ In mathematics, 3 and 4 possess a well-known property: $3^2 + 4^2 = 5^2$, making them the first and most important (due to the fact that 3, 4, and 5 are three consecutive numbers) of the **Pythagorean triplets**.

In addition, 3 is the first non-trivial triangular number, 4 is the first non-trivial square number, and 5 is the first non-trivial interface number (sum of two consecutive squares).

From the perspective of the Torah's inner teachings, triangular, square, and interface numbers are the geometrical equivalents of the three types of interactions found in reality known as, evolution (הִשְׁתַּלְשְׁלוּת), enclothement (הִתְלַבְּשׁוּת), and inspiration (הַשְׁרָאָה). For example, physical objects interacting mechanically exhibit evolutionary interactions. Certain phenomena (such as the soul in the body, and electrical phenomena) demonstrate an enclothement interaction. Finally, phenomena that are non-local are based on an inspirational interaction.

The triplet 3, 4, and 5 also forms the basis of the daily prayers. On weekdays there are 3 prayers known as *Arvit, Shacharit*, and *Minchah*. On Shabbat, we add the *Mussaf*, making it 4 prayers. Finally, on Yom Kippur we add the *Ne'ilah* prayer, bringing the total to 5 prayers.

▶ From the second verse, we learn that the four referred to in the first verse are actually "ways." The words "way" (דֶּרֶךְ) and "thing" (דָּבָר) both have a marital connotation in Torah, as implied by the fourth way, "the way of a man with a maiden" and by another verse, "For he has found in her a lecherous thing..." (Deuteronomy 24:1). Yet, there is a difference, which can be gleaned from the very beginning of *Kidushin*, the Talmudic tractate dealing primarily with betrothal. There we read that, "A woman may be betrothed in three ways" (הָאִשָּׁה נִקְנֵית בְּשָׁלשׁ דְּרָכִים). The ensuing discussion reveals the rationale behind using the word "**ways," rather than** "**things.**" The Talmud states that "way" is used to describe methods that may or may not be applicable in certain circumstances, while the term "thing" appears in a context where such a distinction is unnecessary.

Thus a "way" implies options, requiring choice. Ultimately, this means that the essence of a "way" is rooted deep in the unconscious mind (in the unknowable head of the crown) from where it becomes revealed through choice (as in the case of choosing one's soul-mate).

the most important instance in the Bible where these two numbers are explicitly juxtaposed and associated with one another: two verses written by King Solomon, the wisest of men, who relates to us in Proverbs:[3]

> *There are three [things] that elude me, and four that I know not. The way◀ of the eagle in the sky; the way of a snake upon a rock; the way of a ship in the midst of the sea; and, the way of a man with a maiden.*

שְׁלשָׁה הֵמָּה נִפְלְאוּ מִמֶּנִּי וְאַרְבָּע[ה] לֹא יְדַעְתִּים. דֶּרֶךְ הַנֶּשֶׁר בַּשָּׁמַיִם דֶּרֶךְ נָחָשׁ עֲלֵי צוּר דֶּרֶךְ אֳנִיָּה בְלֶב יָם וְדֶרֶךְ גֶּבֶר בְּעַלְמָה.

First let us notice that, grammatically speaking the word "three" (שְׁלשָׁה) appears in its masculine form, alluding to the three Patriarchs, while the word "four" (אַרְבַּע) appears (in its written form) in the feminine, alluding to the four Matriarchs.▼▼

   Now what about the structure of these verses? Do they too hint at the numbers 3 and 4? In the original Hebrew, the first verse contains 7 words that divide in two. The first part, which refers to the "three," contains 4 words, "Three things elude me" (שְׁלשָׁה הֵמָּה נִפְלְאוּ מִמֶּנִּי). The second part, which refers to the "four" contains 3 words that mean, "and four I know not" (וְאַרְבָּע[ה] לֹא יְדַעְתִּים). So, just the verse's basic structure reflects a pleasing state of both self-reference and inter-inclusion. Self-reference because the two parts of the verse have 4 and 3 words and the verses content refers to 3 and 4. Inter-inclusion

---

3. Proverbs 30:18-19.

▶▶ Since the word "four" is written in its feminine form it follows that relatively speaking, "way" (דֶּרֶךְ) is feminine while "thing" (דָּבָר) is masculine. The value of "way-female" and "thing-male" (דֶּרֶךְ נְקֵבָה דָּבָר זָכָר) is 814, or the filling of **God's Name, Shakai**, שין דלת יוד. This Name corresponds to the *sefirah* of foundation in both the male and female *partzufim*, thus providing a link between both words, "way" and "thing." As we will see later in this chapter, the Name *Shakai*

plays a central role in understanding the relationship of 3 and 4 with the Torah's first verse.

The value of all three words, "way, thing, *Shakai*" (דֶּרֶךְ דָּבָר שׁ־דַּי) is 744, which means that the average value of each word is Abraham (אברהם)—recall that 3 and 4 imply a reference to the Patriarchs and Matriarchs, all of whom stem from Abraham.

in that the references are reversed: the part that refers to 3 has 4 words; the part that refers to 4 has 3 words!▶

The translation of the second verse is, "The way of the eagle in the sky; the way of a snake upon a rock; the way of a ship in the midst of the sea; and, the way of a man with a maiden." In the original Hebrew this verse contains 14 words (double the number of words of the first verse), which divide into four phrases describing the four ways.

The first 2 phrases "The way of the eagle in the sky; the way a snake upon a rock" (דֶּרֶךְ הַנֶּשֶׁר בַּשָּׁמַיִם דֶּרֶךְ נָחָשׁ עֲלֵי צוּר) contain 7 words—3 describing the first way, 4 describing the second.

The 3rd and 4th phrases are, "The way of a ship in the midst of the sea; and, the way of a man with a maiden" (דֶּרֶךְ אֳנִיָּה בְלֶב יָם וְדֶרֶךְ גֶּבֶר בְּעַלְמָה) and they too contain 7 words, but this time 4 words describe the first way and 3 the second!

But, see what happens when we look at the structure of the letters in this verse. Each of the four phrases describing these 4 ways contains exactly 12 letters!

- דֶּרֶךְ הַנֶּשֶׁר בַּשָּׁמַיִם ♦
- דֶּרֶךְ נָחָשׁ עֲלֵי צוּר ♦
- דֶּרֶךְ אֳנִיָּה בְלֶב יָם ♦
- וְדֶרֶךְ גֶּבֶר בְּעַלְמָה ♦

12 as we recall is the product of 3 and 4. Now taking the two verses together, we see that they possess 21 words, where 21 = △6. The first verse contains a variation between its written and spoken forms. The written form (called the *kteev*) has the word "four" in its feminine form (אַרְבַּע), while the spoken form (called the *kree*) has the word "four" in its masculine form (אַרְבָּעָה). Following the spoken form of the word, the first verse has 30 letters. The second verse has 48 letters. And, together, their 78 letters can be arranged in the form of the triangle of 12. Once again, 12 is the product of 3 and 4!

▶ "There are three [things] that elude me, and four that I know not" The word here translated as "elude" (נִפְלְאוּ) literally means "wondrous." In Hebrew, the root form of "wonder" (פֶּלֶא) implies something which is removed or separate from normative consciousness and experience. In the context of this verse in Proverbs, as apparent from the following verse and as explained by the traditional commentaries on the verse, it means that the phenomena to be enumerated appear and immediately disappear without leaving any footprints or evidence of where they were and where they went. Thus, the best way to translate this form of "wonder" is "elude [me]."

In terms of modern physics there is a strong allusion here to **quantum phenomena and the uncertainty principle** at work in nature. One might also contemplate the possible correspondence of the four phenomena enumerated here to the four forces of nature. Three of the four forces—the strong force, the weak force, and the electromagnetic force—are intrinsically connected to one another. The fourth—gravity—remains in a category to itself and is therefore the truly unknowable force, akin to the fourth way, "the way of a man with a maiden." Like gravity that draws two bodies together (it never repulses), male and female gravitate to one another, and this phenomenon is essentially different from the first three, for it leaves an impression, pregnancy [see the *Malbim's* commentary to the verse].

A proverbial triangle

Let us picture these two verses from Proverbs in their triangular form and proceed to analyze it,

```
                    ש
                  ל   ש
                ה   ה   מ
              ה   נ   פ   ל
            א   ו   מ   מ   נ
          י   א   ו   ר   ב   ע
        ה   ל   א   י   ד   ע   ת
      י   ם   ד   ר   ך   ה   נ   ש
    ר   ב   ש   מ   י   ם   ד   ר   ך
  נ   ח   ש   ע   ל   י   צ   ו   ר   ד
ר   ך   א   נ   י   ה   ב   ל   ב   י   ם
ו   ד   ר   ך   ג   ב   ר   ב   ע   ל   מ   ה
```

We first note that the bottom line of the triangle is itself the fourth, final way of the four ways, "And the way of a man with a maiden" (וְדֶרֶךְ גֶּבֶר בְּעַלְמָה). At the triangle's apex we find the letter *shin* (ש), the first letter of the word "three." Together these 13 letters are the top and the bottom of the triangle.

```
                    ש
                  ל   ש
                ה   ה   מ
              ה   נ   פ   ל
            א   ו   מ   מ   נ
          י   א   ו   ר   ב   ע
        ה   ל   א   י   ד   ע   ת
      י   ם   ד   ר   ך   ה   נ   ש
    ר   ב   ש   מ   י   ם   ד   ר   ך
  נ   ח   ש   ע   ל   י   צ   ו   ר   ד
ר   ך   א   נ   י   ה   ב   ל   ב   י   ם
ו   ד   ר   ך   ג   ב   ר   ב   ע   ל   מ   ה
```

Their combined *gematria* is 882, the double square of 21 (the number of words), or 2 · 441, the value of "truth"▼ (אֱמֶת). But, note that the sum of the 7 initial letters of the 7 words in the first verse (שְׁלֹשָׁה הֵמָּה נִפְלְאוּ מִמֶּנִּי וְאַרְבָּעָ[ה] לֹא יְדַעְתִּים) is also 441, or "truth" (אֱמֶת).

The *gematria* of the remaining 65 letters appearing in the 10 lines between the top and bottom of the triangle is 4074, which is a product of 7 and 582. 7 of course is the sum of 3 and 4 (a phenomena we have already seen a couple of times in these verses). 582 is surprisingly none other than the value of the entire base line of the triangle, the

---

▶ The value of **truth** (אֱמֶת) in Hebrew is a perfect square, $441 = 21^2$. All square numbers relate to 4, the first non-trivial square. But, the word "truth" itself, its letters in particular, alludes to 3, the first non-trivial triangular number. The sages state that truth's 3 letters (*alef, mem,* and *tav*) are the first, middle, and final of the Hebrew alphabet. Indeed, *alef* (א) is the alphabet's first letter and *tav* (ת) is its last. But, the Hebrew alphabet has 22 letters, and 22 does not have a midpoint, so how can the letter *mem* (מ) be the middle of the Hebrew alphabet?

There are more than a dozen different answers to this question, with the most straightforward being that this statement refers to the Hebrew alphabet together with the 5 final letters. The order of the alphabet is then,

אבגדהוזחטיכךלמםנןסעפףצץקרשת

Now, the *mem* is the 14th letter, the midpoint of 27.

Truth thus alludes to partitioning the Hebrew alphabet into three equal groups with 9 letters each. *Alef* is the first letter of the first group, *mem* is the middle letter of the second group, and *tav* is the final letter of the third group:

א ב ג ד ה ו ז ח ט
י כ ך ל מ ם נ ן ס
ע פ ף צ ץ ק ר ש ת

The full value of the entire Hebrew alphabet in ordinal numbering is the sum of integers from 1 to 27, which is 378, the value of "electrum" (חַשְׁמַל), perhaps the most mysterious word of the Bible.

The sages explain that this word means, "at times silent, at times speaking" (עִתִּים חָשׁוֹת עִתִּים מְמַלְּלוֹת), the letters' essence. The sum of the first group of letters in ordinal numbering is 45, or the triangle of 9 and the value of "man" (אָדָם). The ordinal value of the letters in the second group is 126, or 9 times 14, the ordinal value of the middle letter mem (מ). The ordinal value of the final group of 9 letters is 207, the value of "light" (אוֹר); literally "electrum" is a synonym for light.

The ordinal values of the three groups of letters—45, 126, and 207—form a linear series whose repeating difference (or base) is 81, the value of the exalted "I" pronoun in Hebrew (אָנֹכִי), the first word of the Ten Commandments.

The tripartite division of the Hebrew alphabet suggested by the word "truth" represents a general rule of creation: that all creatures divide (and then subdivide again and again) into three segments, beginning, middle, and end.

One final point: the normative value of "truth" (אֱמֶת) is 441, or 21^2, the value of, "I will be," the Name *Ekyeh* (אֶהְיֶה), which in the Torah (Exodus 3:14) appears as, "I will be that which I will be" (אֶהְיֶה אֲשֶׁר אֶהְיֶה), suggesting the squaring of 21. But now, calculating the value of "truth" using the method of ordinal numbering described above, we find that it is 42 (א = 1, מ = 14, ת = 27) or 2 times 21, twice the (normative and ordinal) value of "I will be" (אֶהְיֶה), and the (regular) ordinal value of "that which" (אֲשֶׁר).

▶ The *gematria* of "the way of a man with a maiden" (וְדֶרֶךְ גֶּבֶר בְּעַלְמָה) is 582, or 6 times 97, where 97 is the value of **Meheitavel** (מְהֵיטַבְאֵל), the wife of King Hadar, the eighth and most rectified of the kings of Edom (Genesis 36:39). In Kabbalah, the kings of Edom and their reigns symbolize the World of Chaos that precedes the World of Rectification. Of all the kings of Edom, Hadar is the only one whose wife the Torah mentions, and his connection with his wife is interpreted in Kabbalah as indicating that his reign symbolizes the end of chaos and the beginning of rectification, with a state of peace between husband and wife.

In fact, Meheitavel's name itself indicates the inter-inclusion of male and female. The first two letters of her name, מה equal 45, the value of the relatively masculine filling of God's essential Name *Havayah* (יוד הא ואו הא), and the rest of the letters, יטבאל, equal 52, the value of the relatively feminine filling of *Havayah* (יוד הה וו הה).

▶ So (וְדֶרֶךְ גֶּבֶר בְּעַלְמָה)! fourth way, "the way of a man with a maiden" (וְדֶרֶךְ גֶּבֶר בְּעַלְמָה), so we have found how the triangle's base is alluded to in the rest of the triangle.

The six letters down the middle axis of the triangle, ש ה מ י י ה = 370, or 10 (5 plus 5) · 37, the number of Divine lights that shine from God's countenance into creation.[4]

Let us focus on the nine letters that comprise the three corners of three letters in the triangle,

$$
\begin{array}{c}
ש \\
ש\ ל \\
ה\ ה\ מ \\
ה\ נ\ פ\ ל \\
א\ ו\ מ\ מ\ נ \\
י\ ו\ א\ ר\ ב\ ע \\
ה\ ל\ א\ י\ ד\ ע\ ת \\
י\ פ\ ם\ ד\ ר\ ך\ ה\ נ\ ש \\
ר\ ב\ ש\ מ\ י\ ם\ ד\ ר\ ך \\
נ\ ח\ ש\ ע\ ל\ י\ צ\ ו\ ר\ ד \\
ר\ ך\ א\ נ\ י\ ה\ ל\ ב\ י\ ם \\
ו\ ד\ ר\ ך\ ג\ ב\ ר\ ב\ ע\ ל\ מ\ ה
\end{array}
$$

The three letters that comprise the upper corner of the triangle spell the feminine form of the word "three," שָׁלֹשׁ, also the root of "three" in the masculine form as it appears in the verse. Of course, we have here an example of self-reference since the first three letters of the triangle are themselves the root of the word "triangle" (מְשֻׁלָשׁ), in Hebrew. Now, the value of these three letters (שלש) is 630, a product of 3! But, perhaps most beautifully, 630 is itself a *triangular* number: 630 = △35.

The three letters that make up the triangle's lower right corner are רדו, and their sum is 210. But, note that 210 is exactly one-third of 630, the value of the upper triangle and the value of "three" (שָׁלֹשׁ). This means that 210 is the average value of the 3 letters of "three." Once again, 210 itself is a triangular number: 210 = △20.

---

4. See *Eitz Chaim* 4:5.

All nine letters of all three triangles are ש ל ש ר ד ו מ ם ה and their sum is 925, a number with some unique properties. It is both the interface (inspirational) number of 22 ($925 = 22^2 + 21^2$) and the *chashmal* (pentagonal) number of 24 ($925 = \triangle24 + 25^2$). It is also the product of 25 ($=5^2$) and 37. Recall how central 37 is to the first verse of creation.

The final stage of our analysis (for now) involves the triangles periphery,

```
                              ש
                         ש    ל
                    ה  ה  מ
                    ה  נ  פ  ל
               א  ו  מ  מ  נ
               י  ו  א  ר  ב  ע
          ה  ל  א  י  ד  ע  ת
          י  מ  ס  ד  ר  ך  ה  נ  ש
     ר  ב  ש  מ  י  ם  ד  ר  ך
     נ  ח  ש  ע  ל  י  צ  ו  ר  ד
ר  ך  א  נ  י  ה  ב  ל  ב  י  ם
ו  ד  ר  ך  ג  ב  ר  ב  ע  ל  מ  ה
```

The sum of the 33 letters that make up the periphery is 2652, or $26 \cdot 102$. The numbers 26 and 102 have special significance in the Torah as they are the values of *Havayah* (י־הוה), God's essential Name, and "our God" (אֱ־לֹהֵינוּ), respectively. These two words make up the middle phrase of the *Shema*, the statement of Jewish faith in God's oneness, "Hear O' Israel, *Havayah* is our God (י־הוה אֱ־לֹהֵינוּ), *Havayah* is one."

The sum of the remaining 45 letters that make up the inner triangle is $2304 = 48^2$. 48, as you might recall, is the number of letters in the second verse.

But, now, the final point of our present analysis reveals another beautiful phenomenon of self-reference, because both the value of the periphery (2652) and of the inner triangle (2304) are multiples of 12, which we recall is $3 \cdot 4$! When we add both numbers together—

and, thus obtain the full *gematria* of both verses together—we get 4956, which is a multiple of both 7 (3 ⊥ 4) and 12 (3 · 4)!

Analysis through context

There are cases in which the context of a verse or the meaning of its content might suggest a method of mathematical analysis that we would not normally pursue. Since our two verses from Proverbs juxtapose 3 and 4 explicitly, we are prompted to wonder whether a structure of 3 by 4 can be seen within the text.

We have already arranged the 78 letters of the two verses into a triangle of 12 lines. Since 12 = 3 · 4 (and of course also 4 · 3, but we will not pursue this division of 12 at the present), this suggests analyzing the triangle by splitting it into 3 segments with 4 lines in each. Obviously, there is nothing very mathematical or geometric about doing this, but once more, we are prompted to perform this segmentation of the triangle of the two verses because of contextual considerations, not mathematical ones!

Now, if we divide the triangle in this manner, then each of the 4 lines in each segment symbolizes one of the 4 "ways" explicitly appearing in the text.

```
                    ש
                  ש   ל
                ה   ה   מ
              ה   נ   פ   ל
            א   ו   מ   מ   נ
          י   ו   ר   ב   ע
        ה   ל   א   י   ד   ת
      י   מ   ד   ר   ה   נ   ש
    ר   ב   ש   י   מ   ד   ר   ך
  נ   ח   ש   ע   ל   י   צ   ו   ר   ד
ר   ך   א   נ   י   ה   ב   ל   ב   י   ם
ו   ד   ר   ך   ג   ר   ב   ע   ל   מ   ה
```

Now, let us look at these 3 segments more carefully. The first segment (lines 1 through 4) contains 10 letters. The second segment (lines 5 through 8) contains 26 letters. Finally, the third segment (lines 9 through 12) contains 42 letters. This in itself is an important series, where 10 symbolizes the number of *sefirot*, 26 the value of *Havayah*, God's essential Name, and 42 is the number of letters in the Name of 42▸, a holy Name that appears throughout the daily prayers.[5]

But, let us focus on the value of each segment:

• segment 1 comes to 845 (= $5 \cdot 13^2$)
• segment 2 comes to 1575 (= $7 \cdot 15^2$)
• segment 3 comes to 2536▸▸

We can use these three numbers to form a quadratic series, which we can develop forwards and backwards using the method of *finite differences*. First we find the series' base,

$$845 \qquad 1575 \qquad 2536$$
$$730 \qquad 961$$
$$231$$

The base of this series is 231, a number that carries a great deal of significance in Hebrew grammar as it is the number of possible two-letter roots obtained by calculating the value of: $21 \cdot 22/2$. Mathematically, this is the same as calculating the triangle of 21. So 231 is thus the number of possible two-letter combinations of 22 letters, which equals the triangle of 21. 21 itself is a triangular number, $21 = \triangle 6$. And 6 itself is also a triangular number, $6 = \triangle 3$. And, finally, 3 is a triangular number itself with $3 = \triangle 2$. We might say that 231 is *a very triangular number* (the first such number after 1) and we of course recall that we produced this series by segmenting a triangle, albeit in a somewhat untraditional manner.

Using the base, we can now extend the series backwards until its first positive number and forwards until the 7th positive number in the series. We find:

---

5. In the initials of the liturgical poem, אָנָּא בְּכֹחַ.

▸ In their feminine forms, the Hebrew words for "three" (שָׁלֹשׁ) and "four" (אַרְבַּע) refer to themselves, because each is composed of the number of letters it refers to. This is a form of self-reference unique to these two numerals—the names of the remaining numerals are: 1 (אֶחָת), 2 (שְׁתַּיִם), 5 (חָמֵשׁ), 6 (שֵׁשׁ), 7 (שֶׁבַע), 8 (שְׁמוֹנֶה), 9 (תֵּשַׁע). The *gematria* of these two words together, "three" (שָׁלֹשׁ) and "four" (אַרְבַּע) is 903 = $\triangle 42$, a beautiful allusion to the Divine mystical **Name of 42 letters** with which the world was created.

This number, 903, also appears in the Torah's first word, "In the beginning" (בְּרֵאשִׁית). The value of five of its six letters, בראשת, is 903. The value of the sixth letter, the *yud* (י) is 10, also a triangular number: 10 = $\triangle 4$. Thus, the *gematria* of the Torah's first word, "In the beginning" (בְּרֵאשִׁית), 913, is the sum of two triangles, $\triangle 42 \perp \triangle 4$.

Just as 42 alludes to the Divine Name of 42 letters, so 4 alludes to God's essential Name, *Havayah* (יהוה), also called the Tetragrammaton, the Name of 4 letters whose literal meaning is "He who brings all into existence." Extending the connection between the Tetragrammaton and the Name of 42 letters one more step, the value of the first is 26. The value of the latter is 3701, while the value of the source text from which it is generated (the first 42 letters of the Torah) is 3842. Together, all three values equal 7569, a perfect square! $7569 = 87^2$, where 87 is the value of the often repeated declaration, "I am *Havayah*" (אֲנִי יְהוָה)!

▸▸ Adding **2536** to the sum of the first segment 845, we get 3381 = $7^2 \cdot 69$, or $23 \cdot 147$, the value of the last word of the two verses, "with a maiden" (בְּעַלְמָה).

78		346		845		1575		2536		3728		5151
	268		499		730		961		1192		1423	
		231		231		231		231		231		

Now, note that the first positive number in this series is 78—the total number of letters in the two verses and, of course, the triangle of 12!

Analyzing quadratic series is a field to itself in the mathematics of Torah with many interesting results obtainable through straightforward calculations. Let us look at one of them. Before us we have the first 7 positive numbers in this series, with 1575 in the middle. As noted earlier, 1575 is a multiple of 7. Considering the two numbers around 1575—845 and 2536—we see that though neither is divisible by 7, their sum, 3381 equals 7 · 483. It is easy to show algebraically that the sum of every further pair of numbers surrounding 1575 will also be a multiple of 7. Consider the next pair—346 and 3728—their sum is 4074, or 7 · 582. Recall that we saw 4074 and 582 earlier in this chapter.

This also means, according to the general properties of quadratic series, that the sum of these 7 numbers will equal 7 times half of 4074, or, 14259 = 97 · 147, where 97 is once again the value of "Meheitavel" (מְהֵיטַבְאֵל), as noted earlier, and 147 is the value of, "with a maiden" (בְּעַלְמָה), the final word of the two verses!

Proverbs and Genesis together

Now, let us connect these two verses from Proverbs with the Torah's first verse, "In the beginning God created the heavens and the earth" (בְּרֵאשִׁית בָּרָא אֱ־לֹהִים אֵת הַשָּׁמַיִם וְאֵת הָאָרֶץ).

Counting words, we see that the Torah's first verse has 7, which when added to the 21 (=△6) words of the two verses from Proverbs, total 28 words, or △7.

The Torah's first verse has 28 (=△7) letters. When added to the 77 letters of the two verses from Proverbs (when the word, "and four" is spelled in the form it appears in the written text▸) the total number of letters becomes 105, also a triangular number: $105 = \triangle 14$.

To more clearly see what adding the Torah's first verse has done, let us indicate the changes using a table:

	Proverbs	Proverbs with Torah's first verse
words	$21 = \triangle 6$	$(21 + 7 =)\ 28 = \triangle 7$
letters	$78 = \triangle 12$	$(77 + 28 =)\ 105 = \triangle 14$

So, the verses in Proverbs have a triangular number of words and letters, and adding the Torah's first verse to them also gives a triangular number of words and letters. Moreover, adding the Torah's first verse has retained the 1:2 ratio between the indexes of the triangles of letters to words. In other words, the words-to-letters ratio before was $\triangle 6$ to $\triangle 12$ (6 to 12 is a 1:2 ratio) and after adding the Torah's first verse became $\triangle 7$ to $\triangle 14$ (and once again, 7:14 is a 1:2 ratio).

Some algebra and geometry

The fact that the ratio of the triangular indexes of words to letters remained 1:2 is non-trivial and should prompt us to more carefully consider the mathematics behind the result. We should treat it as if the Torah is pointing us in the direction of more rigorously defining a particular mathematical relationship. So let us do just that. Let us look at what happens to a triangular number when its index is doubled.

What we want to look at is the relationship between $\triangle n$ and $\triangle(2n)$. First let us use algebra to see what happens when we double the index.

To do so we must first recall that the general expression for $\triangle n$ is:

$$\triangle n = \frac{n\,(n+1)}{2}$$

▸ We mentioned earlier (p. 93) the difference that sometimes exists in the Biblical text between a word's **spoken (kree) and written (kteev) forms**. This difference, when it appears, represents the Torah giving us a chance to relate to a word, in more than one way. Since it is not pronounced, a word's written form alludes to the word's more concealed dimensions while its spoken form serves to reveal it more literal meanings. In our particular case, the written form has guided us in connecting this verse from Proverbs to a verse from the Pentateuch illustrating one of the most important concepts in studying the Torah's hidden dimensions—"there is nothing that is not alluded to already in Torah [i.e., in the Pentateuch]."

If we now replace n with 2n we get,

$$2n)\triangle) = \frac{2n\,(2n \perp 1)}{2} = 2n^2 \perp n$$

Set n = 6 and n = 7 to convince yourself this expression is correct.

So far, this may not seem very interesting. But, now let us ask, What kind of figurate number does the expression (or function) $2n^2 \perp n$ generate? The simplest answer would of course be a rectangle made up of two squares with side n with one more row of length n. For example, for n = 6, we would get the following rectangle,

We use two different colors of dots in order to highlight the two squares and the additional column. The resulting shape presents an alternate way of drawing the verses we have been looking at. Note that this particular rectangle, since it was drawn based on n = 6, has 78 dots. So the 78 letters of the two verses from Proverbs can be drawn based on this rectangle. Likewise, drawing the rectangle for n = 7, we get,

This rectangle is another way of arranging 105 dots (instead of using the $\triangle 14$ shape, which would be a triangle). We could use this rectangle too as the shape with which to order the 105 letters of the two verses in Proverbs with the Torah's first verse.

A word about rectangles is due. Many verses (and passages) in the Bible can be formed into a rectangular shape. However, for the rectangular shape to bear relevance on the letters' content, we must find justification◄ for using a rectangle in any particular case because

▶ The Torah's first verse has 28 letters, so it can be written in the form of a 7 by 4 rectangle, like so,

ב ת י ש א ר ב
ם י ה ל א א ר
ם י מ ש ה ת א
ץ ר א ה ת א ו

Perhaps the most important phenomenon revealed by doing so is that the middle row of letters, ש ל ש ה, spells the Hebrew word for **three** (שְׁלִשָׁה).

▸ Two additional **exapmles of the rectangular shape** are the Ten Commandments and the Priestly Blessing. The Ten Commandments (in Exodus) contain 620 letters. Rabbi Avraham Abulafia depicts them as a rectangle of 62 columns by 10 rows:

אנכייהוהאלהיךראשרהוצאתיךמאריצרממבריתעבדיםלאהיהלךאלהיםאחריםעלפ
נילאתעשהלךפסללוכלתמונהאשרבשמיםממעלואשרבארץמתחתיאשרבמירסממתחתלארץ
אתתחתוהלהמולאתעבדםכיאנכייהוהאלהיךאלקנאפקדעוןאבתעלבניעלשלשיםו
לרבעישלנשואיעשהחסדלאלפיסלאהביולשמרימצותילאתשאאתשמהיהוהאלהיךלשוא
כילאינקהיהוהאתאשריששאאתשמולשואוכוראתימהשבתלקדשוששתימיםתעבדועשי
תכלמלאכתךויוםהשביעישבתליהוהאלהיךלאתעשהכלמלאכהאתהובנךובתךועבדךוא
מתךובהמתךוגרךאשרבשעריךכיששתימיםעשהיהוהאתהשמיםואתהארץאתהיםואתכל
אשרבםויניחביוםהשביעיעלכןברךיהוהאתיוםהשבתויקדשהוכבדאתאביךואתאמךך
מעןיארכוןימיךעלהאדמהאשריהוהאלהיךנתןלךלאתרצחלאתנאףלאתגנבלאתענהב
רעךעדשקרלאתחמדביתרעךלאתחמדאשתרעךועבדוואמתוושורווחמרווכלאשרלרעך

Being that there are 10 commandments, there is real justification in doing so.

Another example of justified use of a rectangle is found in regards the Priestly Blessing (Numbers 6:24-6), whose 15 words can be ordered as a rectangle of 3 rows by 5 columns.

ייהוה	יָאֵר	וְיִשְׁמְרֶךָ	יהוה	יְבָרֶכְךָ
ייהוה	יִשָּׂא	וִיחֻנֶּךָּ	אֵלֶיךָ	פָּנָיו
שָׁלוֹם	לְךָ	וְיָשֵׂם	אֵלֶיךָ	פָּנָיו

Here too the justification lies in the fact that the Priestly Blessing is actually 3 blessings in one (each blessing appears in a separate verse). Interestingly, in this respect, one of the sacred Names of God derived from the Priestly Blessing is known as the Name of 22 Letters. The logic for selecting 22 letters out of the 60 letters in the full blessing is based on the top row of this rectangular depiction of the blessing: the 5 words appearing in the top row consist of exactly 22 letters.

geometrically (and algebraically) the rectangle is less complex than other figurate number forms▴ (except for the diamond figure, which can also be depicted as an n by $n+1$ rectangle).▸▸

Some rectangles though are more equal than others, so to speak, for example, golden rectangles—rectangles whose two sides are consecutive numbers in the Love (Fibonacci) series of numbers (1, 1, 2, 3, 5, 8, …).

So the question presents itself: can the expression $2n^2 + n$ be depicted in a geometric shape other than a rectangle?

The answer is yes. And the shape that will do the trick is a square of size n with two triangles of size n attached to its top and bottom. The way to arrive at this interesting idea is by noticing that we can rewrite

$$2n^2 + n$$

as simply

$$n^2 + n^2 + n$$

Then all we have to do is notice that $n^2 + n$ is exactly twice as much as the expression for a single triangle of size n, or algebraically,

$$n^2 + n = 2\,\frac{n(n+1)}{2}$$

▸▸ Undoubtedly, the most interesting numbers are the primes, but rectangles cannot be used when depicting a verse or passage possessing a prime number of letters or words, which leads us to develop the entire **theory of figurate numbers**, which do allow us to depict a prime number of elements.

The first series of figurate numbers in which prime numbers appear in significant concentration are the interface, or inspirational numbers (**1**, **5**, **13**, 25, **41**, **61**, 85, **113**, …—numbers in bold are prime). Two other important primes that can be expressed as figurate numbers are 541, the value of "Israel" (יִשְׂרָאֵל), which is a Magen David number, and 613, the number of commandments in the Torah, which is an interface number.

103

So, $n^2 + n$ is actually the number of dots in two triangles of size n! Together with the n^2 we left behind, we have that the $\triangle(2n)$ can be drawn as a square of size n with two triangles of size n.

Let's draw this figure for n = 7:

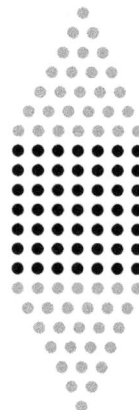

Just like the rectangle we drew earlier, this figure also has exactly 105 dots and we can use it as a model for drawing the 105 letters of our two verses from Proverbs together with the Torah's first verse. Incidentally, we could have drawn the two triangles in this figure on either side of the square, but that would have made for a shape that does not lend itself well to ordering the letters of a text.

For now, we will not carry the analysis of our 105 letters further with this figure, instead we will use the more simple and straightforward shape of the triangle to continue.

The 105 letter triangle

We now proceed to draw the 105 letters that make up the Torah's first verse and our two verses from Proverbs in the shape of a triangle, specifically the triangle of 14. We will highlight the beautiful phenomenon found in this drawing, namely that the 28 letters of the Torah's first verse, "In the beginning God created the heavens and the earth," populate the triangle's first 7 rows, and consequentially, that the two verses from Proverbs populate the next 7 lines,

```
                              ב
                           ר   א
                        ת   י   ש
                     ב   ר   א   א
                  ל   ה   י   ם   א
               ת   ה   ש   מ   י   ם
            ו   א   ת   ה   א   ר   ץ
         ש   ל   ש   ה   ה   מ   ה   ה   נ
      פ   ל   א   ו   מ   מ   נ   י   י   ו
   א   ר   ב   ע   י   א   ל   ע   ד   ת
י   ם   ד   ר   ך   ה   נ   ש   ר   ב   ש
מ   י   ם   ד   ר   ך   נ   ח   ש   ע   ל   י
צ   ו   ר   ד   ר   ך   א   נ   י   ה   ב   ל   ב
י   ם   ו   ד   ר   ך   ג   ב   ר   ב   ע   ל   מ   ה
```

Considering this form of the verses, let us begin its analysis by looking at the 7 letters that run down the middle axis:

```
                              ב
                           ר   א
                        ת   י   ש
                     ב   ר   א   א
                  ל   ה   י   ם   א
               ת   ה   ש   מ   י   ם
            ו   א   ת   ה   א   ר   ץ
         ש   ל   ש   ה   ה   מ   ה   ה   נ
      פ   ל   א   ו   מ   מ   נ   י   י   ו
   א   ר   ב   ע   י   א   ל   ע   ד   ת
י   ם   ד   ר   ך   ה   נ   ש   ר   ב   ש
מ   י   ם   ד   ר   ך   נ   ח   ש   ע   ל   י
צ   ו   ר   ד   ר   ך   א   נ   י   ה   ב   ל   ב
י   ם   ו   ד   ר   ך   ג   ב   ר   ב   ע   ל   מ   ה
```

They are, ב י י ה מ ה א, and their sum is 73, the value of "wisdom" (חָכְמָה). As we have seen above, the world was created with wisdom.

The primordial numbering value of these 7 letters is, 3, 55, 55, 15,

145, 15, and 1 and their primordial sum is therefore 289, or 17^2. We saw in earlier chapters that 289 is also the value of, "God created" (בָּרָא אֱ־לֹהִים), the second and third words of the Torah!

If we take the difference between 289 and 73, we get the value of the additional "primordial letters." One way of seeing this is by writing out the primordial calculation in longhand. For these seven letters, the primordial longhand is: אב, אבגדהוזחטי, אבגדהוזחטי, אבגדה, אבגדהוזחטיכלמ, אבגדהוזחטיכלמ, אבגדה, and א. The additional primordial letters would then be the letters in bold: **אב**, אבגדהוזחטי, **אבגדהוזחטי**, אבגדה, **אבגדהוזחטיכלמ**, אבגדה. Again, their value will be the difference between 289 and 73, which is 216 (= 6^3). 216 is also the value of both the *sefirah* of "might" (גְּבוּרָה) and its inner motivator, "fear" or "awe" (יִרְאָה).

From this we learn that both wisdom and might participate together in the continual recreation of the world, and that this manifests in the soul as the experience of *awe* in the presence of the Divine wonders of creation as they reflect the infinite *wisdom* of the Creator.

Trust in God

Focusing on the letters in the three corners of the triangle,

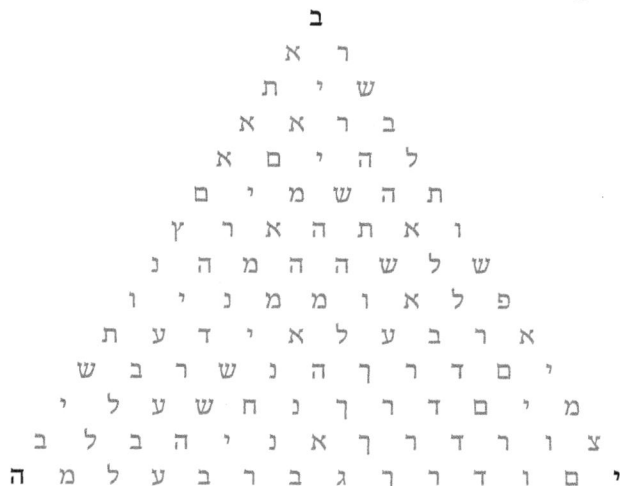

```
                        ב
                     ר  א
                  ת  י  ש
               א  ר  ב  א
            א  מ  י  ה  ל
         ם  י  מ  ש  ה  ת
      ץ  ר  א  ה  ת  א  ו
   נ  ה  מ  ה  ש  ל  ש
ו  י  נ  מ  ו  א  ל  פ
ת  ע  ד  י  א  ל  ע  ב  ר  א
ש  ב  ר  ש  נ  ה  ר  ד  ד  מ  י
י  ל  ע  ש  ח  נ  כ  ר  ד  מ  י  מ
ב  ל  ב  ה  י  נ  א  כ  ר  ד  ר  צ
ה  מ  ל  ע  ב  ר  ג  כ  ר  ד  ו  מ  י
```

We find that these are בֿ י ה, which together spell the word, "in God"

(בְּיָהּ), where God here is represented by the 2 letter shortening of the full Tetregrammaton, pronounced *Kah* (יָהּ). Actually, this word "in God" (בְּיָהּ) appears in one of the most important Biblical verses[6] relating to the mystery of creation: "Trust in *Havayah* for ever and ever, for in God *Havayah* is everlasting strength" (בִּטְחוּ בַּי־הוה עֲדֵי עַד כִּי בְּיָהּ י־הוה צוּר עוֹלָמִים).

How does this verse relate to creation? The last two words of this verse "everlasting strength" (צוּר עוֹלָמִים) are read by the sages[7] as meaning, "[*Havayah*] formed worlds" (צָר עוֹלָמִים). Meaning that the complete phrase reads, "With [the two letters of the Divine Name] *Kah* (יָהּ), God formed [two] worlds [the World to Come with the first letter of *Kah* (*yud*) and this world with the second letter of *Kah* (*hei*)]." The word "with God" (בְּיָהּ) = 17, the square root of "God created" (בָּרָא אֱ־לֹהִים) and the value of the initial letters of "the heavens and the earth" (אֵת הַשָּׁמַיִם וְאֵת הָאָרֶץ), which God created. Incredibly, using the primordial value of each letter, this word, "with God" (בְּיָהּ) equals 73, the value of "wisdom" (חָכְמָה)▸, which you will recall is also the value of the seven letters running down the triangle's middle axis! But, now note (if you haven't already) that the letters that spell "with God" (בְּיָהּ) are also the first, second (and third), and fourth letters down the same axis!

Another beautiful phenomenon is revealed when we look for the word, "with God" (בְּיָהּ) in the first verse of creation. If we set the large letter *beit* of "In the beginning" (בְּרֵאשִׁית) as our starting point in the verse, we find that 2 equal skips of 12 letters spells the word "with God" (בְּיָהּ):

בְּרֵאשִׁית בָּרָא אֱלֹהִים אֵת הַשָּׁמַיִם וְאֵת הָאָרֶץ

Altogether, in the first verse, the word "with God" is encoded over 25 letters—from the *beit* of "In the beginning" (בְּרֵאשִׁית) to the *hei* of "the earth" (הָאָרֶץ). These are actually the first 25 letters in the Torah. So we

---

6. Isaiah 26:4.
7. *Menachot* 29b and *Chagigah* 10a.

▸ One of the more advanced tools in the Torah's mathematical toolbox is the concept of garments, or clothing (מַלְבּוּשׁ). **The garments of a word** consist of the word's three letter-transformations, known as *atbash* (את״בש), *albam* (אלב״ם) and *achbi* (אכב״י) [See our online encyclopedia at www.innerpedia.org for the exact definitions of these transformations]. Of the many transformations used by the sages of Kabbalah, these have a unique property that the three of them together form a transformation ring. What this means is that if you take a letter and run it through all three transformations sequentially in any order, the final output will be the original letter.

The garment of a word is then defined as the sum of all 3 transformations of the word, which are, in the case of "with God" (בְּיָהּ), **יבז משע שמצ**, whose total value is 859. When the original word, "with God" is enclothed within its garment, the total comes to 876. Incredibly, once more this word is seen to relate to wisdom, as the average value of each of the 12 letters is 73, the value of "wisdom" (חָכְמָה).

Another related point: the initial letter of "with God" (בְּיָהּ), the letter *beit* (ב) in its garment, is בשמי, which is a word that means, "in My Name" (בִּשְׁמִי). This word appears in another central Biblical verse describing creation, "All that is called in My Name (בִשְׁמִי) and for My glory, I have created it, I have formed it, even have I made it" (Isaiah 43:7).

have here an interesting connection between 12 (the size of the skip) and 25 (the total number of letters included from beginning to end of the encoding). What makes this connection so interesting is that 12 and 25 are at the heart of the difference between Moses' prophecy and the prophecy of all other true prophets. The sages explain that the clarity of the prophecy experienced by all true prophets can be likened to a person looking through a translucent pane of glass,[8] but Moses' prophecy is like a person looking through a transparent pane of glass.[9] This is seen by the initial word each uses to describe their prophecy. All prophets but Moses begin with the word, "So [says God]" (כה), whose value is 25. Moses begins his special prophecies with the word, "This [is what God said]" (זה), whose value is 12.

Finally, as we did with the previous triangle, let us focus on the three-letter corners,

```
                                    ב
                                  ר   א
                              ש   י   ת
                          ב   ר   א   א
                      ל   ה   י   ם   א
                  ת   ה   ש   מ   י   ם
              ו   א   ה   ת   ה   א   ר   ץ
          ש   ל   ש   ה   ה   מ   ה   נ
      פ   ל   א   ו   מ   מ   נ   י   ו
  א   ר   ב   ע   ל   א   י   ד   ע   ת
ש   ם   ד   ר   ך   ה   נ   ש   ר   ב   ש
י   מ   ם   ד   ר   ך   נ   ח   ש   ע   ל   י
צ   ו   ר   ד   ר   ך   א   נ   י   ה   ב   ל   ב
י   ם   ו   ד   ר   ך   ג   ב   ר   ב   ע   ל   מ   ה
```

The nine letters in these corners are ברא צים במה and their total sum is 390, which beautifully refers back to the earlier phenomena, because 390 = 15 · 26, where 15 is the value of God's Name, Kah (יה) and 26 is of course the value of Havayah (יהוה)! Again, this is an allusion to the verse, "Trust in Havayah for ever and ever, for in God Havayah

8. *Yevamot* 49b.
9. See also Maimonides' commentary on the Mishnah, *Keilim* 30:2.

is everlasting strength" (בְּטְחוּ בַי־הוה עֲדֵי עַד כִּי בְּיָ־הּ י־הוה צוּר עוֹלָמִים), since the words "God *Havayah*" are actually these two Names, *Kah* and *Havayah*.

Let us go one step further by looking at the 6-letter corners,

```
                              ב
                           ר    א
                        ש    י    ת
                     א    ר    ב
                  א    י    ה    ל
               ם    י    מ    ש    ה    ת
            ץ    ר    א    ה    ת    א    ו
         נ    ה    מ    ה    ש    ל    ש
      ו    י    נ    מ    מ    ו    א    ל    פ
   ת    ע    ד    י    א    ל    ע    ב    ר    א
ש    ב    ר    ש    נ    ה    כ    ר    ד    מ    י
י    ל    ע    ש    ח    נ    כ    ר    ד    מ    י    מ
ב    ל    ב    ה    י    נ    א    כ    ר    ד    ר    ו    צ
ה    מ    ל    ע    ב    ר    ג    כ    ר    ד    ו    מ    י
```

The eighteen letters marked are בראשית מצוימו ילבלמה and their sum is 1222 = 47 · 26. First of all, note that if we add these two factors of 1222, we get 47 ⊥ 26 = 73, the value of "wisdom" (חָכְמָה). Another way of expressing 1222 is 2 · 611, where 611 is the value of "Torah" (תּוֹרָה), alluding to the Written Torah and the Oral Torah.

Abulafia's mysterious phrase

Strengthening the already well-developed connection between the numbers 3, 4, and the Bible's first verse, we find a unique *gematria* introduced by the great Kabbalist, Rabbi Avraham Abulafia. He writes[10] that the *gematria* of "In the beginning" (בְּרֵאשִׁית) 913, is the same as that of the phrase "three things together" (שְׁלֹשָׁה דְּבָרִים יַחַד).[11] These three things allude▶ to the three things implied in the first

▶ What Abulafia was actually referring to is that the Hebrew language is based on **3-letter verbs** from which all the words in the language are derived. Each 3-letter verb can be permuted in 6 different ways. In fact the word for "permutation" (צֵרוּף) in Hebrew is a synonym for "together" (יַחַד). In chapter 8, we will see how the Torah's first word, "In the beginning" (בְּרֵאשִׁית) already suggests the 6 possible permutations of 3 letters.

---

10. *Chayei Ha'olam Haba*, p. 113.
11. See also ch. 2, p. 30.

▶ With a small change in vocalization, the phrase, "Three things together" (שְׁלֹשָׁה דְּבָרִים יַחַד) can be read as, "**Three leaders** (דָּבָרִים) together." Note that this alternate rendering does not change the phrase's numerical value.

The value of just the first two words, "Three leaders" (שְׁלֹשָׁה דָּבָרִים) is 891, which incredibly is also the numerical value of the names of the three sibling-leaders: Moses (מֹשֶׁה), Aaron (אַהֲרֹן), and Miriam (מִרְיָם), whose *gematria* are 345, 256, and 290, respectively. These three leaders together (see Shabbat 88a) redeemed us from Egypt and served as the spiritual conduits to convey the word of God in the Giving of the Torah at Mt. Sinai.

Between them, Moses, Aaron, and Miriam's names possess 11 letters. Since 891 = 81 · 11, the average value of each letter is 81 = 9^2 = 3^4. In a beautiful example of self-reference, 81 is also the value of the first and most essential word of the Ten Commandments, the exalted "I" (אָנֹכִי), the word expressing God's very essence.

The word "leaders" (דָּבָרִים) itself is equal to 256 (= 16^2 = 4^4 = 2^8), the *gematria* of "Aaron" (אַהֲרֹן), the middle sibling. It follows then that the value of the word "three" (שְׁלֹשָׁה), appearing in the phrase, "Three leaders," which is 635, is equal to that of "Moses" (מֹשֶׁה) and "Miriam" (מִרְיָם), 345 and 290, the other two sibling-leaders.

▶▶ Inter-inclusion is related to **square numbers**. If there are 3 things being inter-included, the number of properties revealed would be 3^2 or 9. In general, any number of n things inter-included yields n^2 properties. This is one of main reasons we identify square numbers with true unity or complementarity.

verse from Proverbs that we have been looking at, "There are three [things] that elude me, and four that I know not."

To gain a better appreciation of this special phrase and its significance requires some analysis.

Let us begin by dividing the phrase into two parts, "three" and "things together." We are motivated to do so because obviously, if "three four" (שְׁלֹשָׁה אַרְבָּעָה) equals "three things together" (שְׁלֹשָׁה דְּבָרִים יַחַד) then just the word "four" (אַרְבָּעָה) has the same value as "things together" (דְּבָרִים יַחַד).

Now, the sages teach that "the minimal plurality is two"[12] (מִעוּט רַבִּים, שְׁנַיִם) meaning that when a noun appears in the Torah in its plural form, but without an explicit indication of an amount, then one may assume that the plurality refers to two of that noun. In our case, when taking the second part of the phrase as independent from the first, the word "things" (דְּבָרִים)◀ appears in its plural form but is no longer explicitly defined by "three." Therefore, we can assume that it refers to two different things.

But, when we add the word "together" (things together) we learn that the two things referred to by "things" must come together. True togetherness implies real unity, which in Torah is described as inter-inclusion. Inter-inclusion is more than just a summation or additive joining of two things (be they numbers or any other objects). Inter-inclusion reflects a state of unity created when each of two (or more) things reveals itself in the other(s). For instance, when a man and woman come together in marriage, their union reveals not just their essential masculinity and femininity, but a dimension of their opposite that exists within them. Thus, male and female exhibit 4 characteristic dimensions in marriage: masculine, the feminine within the masculine, the masculine within the feminine, and the feminine. Real togetherness results in the husband revealing his feminine side and the wife revealing her own masculine side. So, with inter-inclusion◀◀, genuine togetherness, the joining of 1 and

---

12. See *Ketubot* 7a.

1 actually yields 4. Finally, we thus have that "things together" actually alludes to the number 4.

But, what this means is that the *gematria* of Abulafia's mysterious phrase, "Three things together" (שְׁלֹשָׁה דְּבָרִים יַחַד) not only equals the words "three four" (שְׁלֹשָׁה אַרְבָּעָה) it actually implies the same idea.

If we continue to consider just the second part of the phrase, "things together," we find that it also equals the value of an important term the sages relate to the creation of light on the first day, "the hidden light" (אוֹר הַגָּנוּז). The identity between the number "four" and "the hidden light" helps us understand why it is that King Solomon relates that, "Four, I do not know." Incredibly, when we write the term "the hidden light" (אוֹר הַגָּנוּז) in full, ▶ אלף ואו ריש הה גימל נון ואו זין, we find that its *gematria* is 913, identical with the *gematria* of the Torah's first word, "In the beginning" (בְּרֵאשִׁית). But, since the initials (called the root letters in the case of filling) equal the value of "four," or 278, this means that the filling letters of "the hidden light," equal the remainder from 913, or "three" (שְׁלֹשָׁה), 635!▶▶

Limiting the infinite

When Rabbi Avraham Abulafia discusses his mystery phrase, "Three things together," he notes that its initial letters spell God's Name pronounced *Shakai* (שַׁדַּי) and usually translated as "the Almighty."▼▼▼ The sages explain that the Name *Shakai's* three letters serve as shorthand for the words, "He who said [to His world],

▶ **The hidden light** (אוֹר הַגָּנוּז) refers to the light created on the first day of creation and concealed by God. It not only refers to the number 4, it also alludes to the juxtaposition of 3 and 4. The filling of "light" (אוֹר) appearing in the text is אלף ואו ריש, which equals 634. When adding the inclusive one, known as the *kollel* (הַכּוֹלֵל), it becomes equal to the value of "three" (שְׁלֹשָׁה), 635. The filling of "hidden" (הַגָּנוּז) used is הה גימל נון ואו זין, which equals 279, also the value of "four" (אַרְבָּעָה), plus the inclusive one. The "inclusive one" represents the energy connecting the letters of a word together, making them one; adding it to the value of a word is ubiquitous in Kabbalistic writings.

▶▶ Note that the filling used here combines both the *mah* (מה) filling for the letter *vav* (ואו) and the *ban* (בן) filling for the letter *hei* (הה).

▶▶▶ Recall that earlier we delved into the relationship between the Name *Shakai*, ways, and things. The letters of *Shakai* (שַׁדַּי) can also stand for, "three ways together" (שְׁלֹשׁ דְּרָכִים יַחַד), whose value is 926. The sum of "three things together" (שְׁלֹשָׁה דְּבָרִים יַחַד) and "three ways together" is then 913 ⊹ 926 = **1839**, a famous number in Kabbalah. 1839 is the value of the verse, "Open my eyes and I will see the wonders of your Torah" (גַּל עֵינַי וְאַבִּיטָה נִפְלָאוֹת מִתּוֹרָתֶךָ). It is also the product of 3 times 613 and the value of "Torah, prayer, repentance" (תּוֹרָה תְּפִלָּה תְּשׁוּבָה), three words that begin with the letter *tav* (ת) whose value is 400, providing another juxtaposition between 3 and 4.

Still, in the verses from Proverbs, we saw that there are four ways, thus implying that the phrase should be "four ways together" (אַרְבַּע דְּרָכִים יַחַד), but then the initials would no longer spell *Shakai* (שַׁדַּי). One way then of understanding the meaning of the phrase "three ways together" (which would retain the initials *Shakai*) is that though the Matriarchs are four—Sarah, Rivkah, Rachel, and Leah—Rachel and Leah were sisters and both wed to the same patriarch, Jacob. Thus, 4 (Matriarchs) can be included in 3 (Patriarchs). From a more essential perspective, in the future, the personas of both Leah and Rachel will be fused into one persona called, Greater Rachel (רָחֵל הַגְּדוֹלָה).

▶ The three words in the phrase, "**He who said Enough!**" (שֶׁאָמַר לְעוֹלָמוֹ דַי) are equal in value to 541, 182, and 14, respectively. These are also the values of "Israel" (יִשְׂרָאֵל), "Yaakov" (יַעֲקֹב), and "David" (דָוִד). It follows then that the value of the entire phrase is 737, also the *gematria* of "a flame" (שַׁלְהֶבֶת) and of the words (Deuteronomy 6:5), "[You shall love God] with all of your heart, with all of your soul and with all of your might" (וְאָהַבְתָּ אֵת הוי' אֱ־לֹהֶיךָ בְּכָל לְבָבְךָ וּבְכָל נַפְשְׁךָ וּבְכָל מְאֹדֶךָ).

'Enough!'" (שֶׁאָמַר [לְעוֹלָמוֹ] דַי)◀. Thus, it is with this Name that God limits creation, by saying, "Enough!" Otherwise, the nature of the Working of Creation is to expand infinitely.[13]

Thanks to General Relativity (and despite Einstein's attempt to introduce a constant that would allow for a static universe), the scientific meaning of an expanding universe is familiar. But, what does it refer to in Torah? Simply put, creation's expansion is a metaphor for the distance that consciousness travels from being one with Divine consciousness. The most far-removed consciousness feels that it is its own beginning and its own end; it came into being at some point in time and will step out of being at another, it has no origin and it is going nowhere. This is a perfectly good description of the nihilistic consciousness, which allows for a life absolutely removed from any responsibility towards the Creator or creation. To prevent this, to guard creation and its creatures from completely dislodging from Divine consciousness and thereby disassociating from the Creator, God uses the Name *Shakai* to impose a limit on creation's expansion. For this reason *Shakai* is also called the Name of the "covenant" as it maintains our ability to uphold a conscious covenant or connection with the Creator and His purpose in creating reality.

Shakai, three, and four

The Name *Shakai* (שַׁ־דָי) has 3 letters and its middle letter is *dalet* (ד), whose value is 4, clearly illustrating its role as the unifier of 3 and 4.

The Name *Shakai* with the phrase, "three things together," reflect the unification of 3 and 4 in a functional manner as well. In Kabbalah, the Name *Shakai*, "the Almighty" is associated with the *sefirah* of foundation. The point occupied by foundation on the Tree of Life of the *sefirot* is where the three axes—right, left, and middle—join. The three axes represent the three primary emotive *sefirot* that lie along them, loving-kindness, might, and compassion. In other words,

---

13. *Chagigah* 12a.

when a reference is made to the "left axis" of the *sefirot*, the essential quality referred to is might; when reference is made to the "right axis," a reference is being made to loving-kindness; likewise for the middle axis, which refers to compassion.

The most powerful image associated with the number 3, the 3 Patriarchs—Abraham, Isaac, and Jacob—mentioned earlier in this chapter, are considered the archetypal souls of loving-kindness, might, and compassion. Thus, foundation is the *sefirah* where the spiritual origins of the souls of the three Patriarchs, Abraham, Isaac, and Jacob come together. It is very fitting then that the Name *Shakai*—the Name associated with foundation—forms the initials of the phrase, "Three things together." The three axes of the *sefirot* coming together at foundation also reminds us of the shape of the letter *shin* (שׁ) whose 3 stems correspond to the 3 Patriarchs, and for that reason is known as "the letter of the Patriarchs."

The light and energy moving down from all three axes fills the vessel of foundation. The *sefirah* of foundation then conveys the light and energy of the 3 axes to the final *sefirah* of kingdom, the feminine principle.▸ Kingdom is the spiritual origin of the souls of the 4 Matriarchs, Sarah, Rebeccah, Leah, and Rachel and is oftentimes referred to as "the fourth." The *sefirah* of foundation, which lies at the junction between the 3 axes (represented by the 3 Patriarchs), also feeds the spiritual source of the 4 Matriarchs. We have found how the Name *Shakai* (שַׁ־דִי), "the Almighty," clearly alludes in another way to the union of three and four.

Let us return to the two phrases that sparked our interest in foundation: "Three things together" (שְׁלֹשָׁה דְּבָרִים יַחַד) and "He said to His world, 'Enough'" (שֶׁאָמַר לְעוֹלְמוֹ דָּי).

Each phrase possesses exactly 3 words and 12 letters. So the number of letters, 12, is already the product of 3 and 4. Also note that between them, the two phrases contain 3 letters *dalet* (ד), and since the value of *dalet* is 4, this again alludes to the union of 3 and 4.

▸ The *sefirah* of kingdom, the feminine principle, is the fourth leg of the **Divine Chariot**, the first three legs corresponding to the three primary emotive *sefirot* of loving-kindness, might, and beauty (the soul source of the three Patriarchs, Abraham, Isaac, and Jacob).

113

SIX IN CREATION

We began the previous chapter with the question of what number lies at the heart of creation. Our answer, three and four together was certainly surprising. This chapter begins with the question of what was the first number to be created once creation got underway.

This may seem a strange question to ask, because how could any number precede the number 1? But, we are not asking a mathematical question (about what the order of the numbers is), rather our focus in on creation's form and function. Is there a particular number that more than any others, lends itself to the model, the form, with which God created reality?

Without a doubt, the first number that should come to mind is the number 6, because God used it when defining the super-structure of creation▶. The world was created in 6 days,▼▼ from the creation of light on the first day to the creation of man, the pinnacle of creation, on the sixth day. And, on the seventh day God rested from creation,

▶ In nature, the number 6 is intrinsically associated with life, since it is **the atomic number of carbon**, the element that forms the basis of all life as we know it.

The ancient world, including the Rabbinic sages of the *Mishnah* and *Talmud*, recognized 4 basic elements: fire, air, water, and earth. Corresponding these to our more modern understanding of the elements, we find that fire corresponds to Carbon, air to Oxygen, water to Hydrogen, and earth to Nitrogen, which (together with Phosphorous and Sulfur) constitute the basic building blocks of life—6 foundational elements in all. As we shall see in chapter 9, the Sages speak of 6 forms of elemental fire.

In nature, the purest allotropes of Carbon, the 6th element, such as diamond and graphite, are found as crystalline structures joined in a 6 pointed lattice sometimes referred to as a honeycomb.

▶▶ In Judaism, the number six is inherently associated with the Oral Torah, specifically **the Mishnah**, whose structure is known by the phrase, "six orders of Mishnah" (שִׁשָּׁה סִדְרֵי מִשְׁנָה). This association has many numerical examples.

First, the value of "six" (שִׁשָּׁה)—in the masculine form, since the word "orders" (סִדְרֵי) is masculine—is 605, the value of the phrase, "A law given to Moses at Sinai" (הֲלָכָה לְמֹשֶׁה מִסִּינַי). This phrase, coined by the earliest sages, is the foundation of the Oral Torah. The value of "six" plus "Mishnah" (מִשְׁנָה) is 1000, an allusion to the 1000 additional lights given to Moses at Sinai, as discussed in the Arizal. The full phrase, "Six orders of Mishnah"—appearing most importantly in the liturgical poem, "Who knows one?" (אֶחָד מִי יוֹדֵעַ)—equals 1274, the product of 49 times 26, the value of God's essential Name, *Havayah*. In this same poem, the full phrase on the Mishnah is, "Who knows six? I know six! Six orders of Mishnah" (שִׁשָּׁה מִי יוֹדֵעַ? שִׁשָּׁה אֲנִי יוֹדֵעַ! שִׁשָּׁה סִדְרֵי מִשְׁנָה), whose value is 2775. We are already intimately familiar

with the number 2701 (the value of the Torah's first verse) and know that it is the triangle of 73. It is easy to see that 2775 is 2701 + 74, meaning that it is the triangle of 74, where 74 is the value of the word "witness" (עֵד). This word appears in the verse (1 Samuel 12:5), "God is witness in you." Thus, whereas the triangle of 73 and the Torah's first verse refer to the creation of Worlds, the triangle of 74 and the number 6 relate to the creation of Souls—the souls of the Jewish people imbued with the Presence of the Creator. Indeed, the number of soul roots of the Jewish people is 600,000, another relationship to 6 (also note that Adam was created on the 6th day of creation).

The value of "6, 60, 600, 6000, 60000, 600000" (שֵׁשׁ שִׁשִּׁים שֵׁשׁ מֵאוֹת שֵׁשֶׁת אֲלָפִים שִׁשִּׁים אֶלֶף שֵׁשׁ מֵאוֹת אֶלֶף)—6 multiplied by 6 successive powers of 10, from 10^0 to 10^5—is $5382 = 207 \cdot 26$, where 207 is the value of "light" (אוֹר) and 26, again is *Havayah* (יהוה)—the light of *Havayah*. Incidentally, all these forms of 6 appear in the Bible.

▶ The form in which the word "six" appears in this verse is שֵׁשֶׁת, pronounced **Sheshet**. There is a well-known Talmudic figure from Babylon whose name was Rav Sheshet. The value of his name with title (רַב שֵׁשֶׁת) is 1202, exactly the value of the Torah's first three words, "In the beginning God created" (בְּרֵאשִׁית בָּרָא אֱ־לֹהִים).

The value of just "six" in this form (שֵׁשֶׁת) is 1000, the same as the value of "six Mishnah" (שִׁשָּׁה מִשְׁנָה), alluding to the six "orders" that make up the *Mishnah*.

▶▶ The sum of **the first two 6-letter words** in the Torah, בְּרֵאשִׁית and וּלְמִקְוֵה, is 1100, exactly the value of "with six days" (שֵׁשֶׁת יָמִים)!

In ordinal numbering (מִסְפָּר סִדּוּרִי), the sum of these two words is 137—the inverse of the fine-structure constant in physics, and thus one of the secrets of creation. 137 is also the reduced value of the entire first verse of the Ten Commandments (אָנֹכִי יְ־הוָה אֱ־לֹהֶיךָ אֲשֶׁר הוֹצֵאתִיךָ מֵאֶרֶץ מִצְרַיִם מִבֵּית עֲבָדִים).

The difference between their normative value and their primordial value is 1100 - 137 = 963, which is also the value of the phrase (Genesis 1:4), "God saw that the light was good" (וַיַּרְא אֱ־לֹהִים אֶת הָאוֹר כִּי טוֹב). Incredibly, 963 also equal 9 · 107, where 107 is once again the distance between the two words!

or put another, more informative way, on the seventh day God endowed creation with the experience of rest.[1]

This very clear division is exactly how creation is described in the Ten Commandments,[2]

> Six days you shall toil and complete all your labor. And the seventh day is Sabbath, for Havayah your God; you shall not perform any labor…. Because, **for six days** God made the heavens and the earth… and He rested on the seventh day.

The identification of six as the form of creation, the mold as it were, with which God creates and recreates the universe, is found in these words, "for six days God made the heavens and the earth." Because of the particular grammatical form of the words "six days" ◀ (שֵׁשֶׁת יָמִים), a more accurate translation would be, "with six-days." Normally, the masculine form of "six" ("days" in Hebrew has masculine gender) is שִׁשָּׁה. But, here the number six appears in its unifying form (הַטָּיָה שִׁתּוּפִית), indicating that the six days are a single unit, again a type of form or mold that repeats, like a fractal form, throughout creation.

The pre-conscious six

One need not finish the entire first chapter of Genesis to discover that the form of creation lies in the number six. Actually, the very first word of the Torah, "In the beginning" (בְּרֵאשִׁית) tells the whole story. The first thing to notice is that it contains 6 letters. A word with 6 letters is not a common occurrence for the Torah. In fact, one must advance another 107 words to reach the second word that possesses 6 letters, "And to the gathering of" (וּלְמִקְוֵה). ◀◀

But, it is not only the number of letters in the Torah's first word that alludes to the number 6. The sages[3] divide the 6 letters of "In

---

1. See *Rashi* to Genesis 2:2.
2. Exodus 20:9-11. See also Ibid. 31:17.
3. *Zohar* I, 3b.

the beginning" into 2 three-letter words, reading "[He] created six" (בְּרָא שִׁית), referring to all that God created in six days. Dividing a word in two in Hebrew is a common method for understanding its meaning.▸ Moreover, as revealed by the Torah's mystical tradition, the Torah's text in its eternal form, as it stands before God, is not divided into separate words.[4] When Moses received the Torah, he was also told by the Almighty how to divide the long string of letters into words and sentences.

Still, when dividing the six letters of "In the beginning" into these two words, "[He] created six," the word for "six" here (שִׁית) is the Aramaic word, not its Hebrew counterpart (שִׁשָׁה). This point is meaningful because Aramaic is considered Hebrew's posterior (אֲחוֹרַיִּים), or subconscious sister-language, implying that the number 6 is still relatively not consciously present at the very beginning of creation. We will presently see that the six relates to the six *alef*s in the Torah's first verse. The world was created with a *beit*, the first letter of, "In the beginning" (בְּרֵאשִׁית), and so the *beit*, the second letter of the *alefbeit* alludes to the consciousness of creation, whereas the *alef* that precedes the *beit* in the order of the *alefbeit* alludes to the unconscious side of creation. At the moment of the beginning of creation, the six days to come were still lying in a pre-conscious state vis à vis reality.

Permutations of creation

Another way of interpreting the division of the first word into the two words, "created" (בְּרָא) and "six" (שִׁית) is to note that "created" has three letters, which can be permuted in 6 different ways. One of the foundational topics in Torah mathematics is combinatorics. The first book of Kabbalah (and of Hebrew grammar as well), *Sefer Yetzirah* includes the following statement,

> *One stone builds one house; two stones build three houses;*

▸ There are many examples of **compound words in the Bible**. One such word in Genesis 31:47 is *gal'ed* (גַּלְעֵד), meaning a "stone monument." The Bible's most mysterious word, *chashmal* (חַשְׁמַל), appearing in Ezekiel 1:4 is explained by the sages as a compound verb, indicating alternating silence and speech.

---

4. See Ramban's introduction to Genesis.

three stones build six houses; four stones build twenty-four houses; five stones build one-hundred and twenty houses; six stones build seven-hundred and twenty houses; seven stones build five thousand and forty houses...

The "stone" in this passage stands for a letter and the "house" built out of stones is symbolic of a word◂. Thus, as is known in combinatorics, the number of possible permutations of n objects is n! (read: n factorial), the product of all the integers up to n. Thus, the number of possible 3 letter words produced by 3 different letters is $1 \cdot 2 \cdot 3 = 6$.

Now, since the word "created" (בָּרָא) is composed of three different letters, it can be permuted to generate (or build) six different words-permutations. So the first word of the Torah, "In the beginning" (בְּרֵאשִׁית), suggests the combinatoric equality that the first three letters, בְּרָא, can be combined in "six," שִׁית, different ways.

This interpretation is further supported by the fact that the next word after "In the beginning" (בְּרֵאשִׁית) is itself "created" (בָּרָא), the first of the six permutations. It is followed by another 5 words in the first verse, each of which can be symbolically identified with one of the 5 remaining permutations. From various Kabbalistic sources, we learn how to order the permutations between themselves. According to Rabbi Avraham Abulafia, the algorithm is: reverse the order of as few letters from the end of the word so as not to return to a previous permutation. According to this rule, beginning with abc the order of the six permutations is: abc, acb, bca, bac, cab, cba. The last permutation is always the reverse order of the first. Given the three letter ברא, the order of the permutations would then be: ברא, ראב, רבא, אבר, ארב.

A most perfect number

In the previous chapter, we saw that the Torah's first word, "In the beginning" (בְּרֵאשִׁית) equals the phrase "three things together" (שְׁלֹשָׁה

▶ The sum of the numerical values of the two words "**stones**" (אֲבָנִים), 103, **and** "**houses**" (בָּתִּים), 452, is 555, which is a multiple of 37 (specifically: $555 = 37 \cdot 15$). The value of the entire phrase, "three stones build six houses" (שָׁלֹשׁ אֲבָנִים בּוֹנוֹת שִׁשָּׁה בָּתִּים) is 2254, a multiple of 23, the companion number of 37. In the main text, we will see how 23 and 37 appear yet again in another illustration of the number six in creation. Specifically, $2254 = 23 \cdot 98$, the latter being the double-square number of 7. Double-square numbers play a special role in creation, since they have a prominent role in the mathematical structure of the Periodic Table of the elements.

דְּבָרִים יַחַד). Indeed, Rabbi Avraham Abulafia explicitly interprets this equality as suggesting that the first word of the Torah hints to the number of ways in which 3 things can be "together," which is of course 6. The number 6 is referred to as a perfect number▸, meaning that the sum of its proper divisors (it's divisors, not including itself) is itself. Its proper divisors are 1, 2, and 3, and their sum 1 ⊥ 2 ⊥ 3 = 6.

But, 6 is unique in that the product of its proper divisors too is 6!▸▸

These two special properties of the number 6 led Rabbi Avraham Ibn Ezra, one of the greatest early commentaries on the Torah to declare it, the most perfect of all numbers.

6 millennia

The number six appears in another most beautiful phenomenon in the Torah's first verse. Looking at the different letters making up the verse, we find as noted above that the letter *alef* (א), the first letter of the Hebrew alphabet, appears a disproportionate 6 times. Moreover, these 6 appearances of the letter divide evenly between the verse's two halves:

בְּרֵאשִׁית בָּרָא אֱ־לֹהִים אֵת הַשָּׁמַיִם וְאֵת הָאָרֶץ

One of the meanings of the name of the letter *alef* (אָלֶף) is "thousand" (אֶלֶף). The Psalmist says, "For a thousand years in Your eyes are like yesterday gone by."[5] Based on the comparison of a thousand years to a single day in God's eyes, we can learn that the 6 *alef*s in the Torah's first verse allude to the six millennia of this world order (which themselves correspond to the six days of creation), as the sages say, "the world has six thousand years."[6]

Apart from the six *alef*s the Torah's first verse contains 22 letters, whose value is 2695 (note that the single *vav* appearing in the verse is the 22nd letter). 2695 = 49 · 55, where both 49 and 55 are significant figurate numbers, 49 being the square of 7 and 55 the triangle of 10.

---

5. Psalms 90:4.
6. *Sanhedrin* 97a.

▸ Numbers whose proper divisors' sum is the number itself are called **Perfect Numbers**. The first 6 perfect numbers are 1, 6, 28, 496, 8128, and 33550336. All perfect numbers can be found by taking the triangle of certain primes, defined as those primes that are the sum of powers of 2, from 2^0 to 2^m. In fact, the sum of these powers of 2 yields primes known as Merssenne primes in mathematics. They are all equal to $2^{m+1} - 1$. What this means is that the first 6 perfect numbers are the triangles of 1, 3, 7, 31, 127, 8191.

The Torah's first verse actually includes very clear allusions to the first 3 perfect numbers. 6 is the first non-trivial or significant perfect number. It is, as we have seen, the number of letters of the Torah's first word. The major *beit* with which the Torah's first word begins, stands apart, and refers to the number 1, the first perfect number. As we have seen, the first verse has 28 letters, the third perfect number.

Interestingly, the sum of the first 4 non-trivial perfect numbers—6, 28, 496, and 8128—is 8658 = 13 · 666, where 666 has three 6's and is also the triangle of the square of 6. 666 = △(6^2).

▸▸ The sum of 1, 2, and 3 is the "triangle of 3" (△3). The product of 1, 2, and 3 is called "3 factorial" (3!). The ratio of 6 to 3 is 2:1, the secret of "whole and half" (שָׁלֵם וְחֵצִי).

In addition, 2695 is equal to the product of 7 and 385, which means that the average value of the 7 words excluding the 6 *alef*s is 385. 385 is the pyramid of 10.

We define the pyramid of n as:

$$\text{🔲}n = \sum_{k=-1}^{n} k^2$$

Meaning that the pyramid of n is the sum of all squares from 1^2 to n^2. In our case,

$$\text{🔲}10 = 1^2 + 2^2 + 3^2 + 4^2 + 5^2 + 6^2 + 7^2 + 8^2 + 9^2 + 10^2$$

Actually, the calculation of the sum of squares from 1 to 100 is one of the most sophisticated calculations found in the Tikunei *Zohar*,[7] since 385 is also the value of the "Divine Presence" (שְׁכִינָה). Indeed, it is fitting that the Divine Presence—associated with the *sefirah* of kingdom—is equal to the square-pyramid of 10, because kingdom is the 10th *sefirah* and receives the light from all 10 *sefirot* inter-included. ◄

Positional values relative to the verse

One of the most revealing analyses regarding the Torah's text has to do with the position of unique phenomena in the verses being analyzed. The positional values of the 6 *alef*s among the 28 letters of the verse are,

3, 9, 10, 15, 23, 26

Looking at the positional values, we see that the sum of the first four—3, 9, 10, and 15—is 37. The next positional value, 23, is considered the companion number of 37 (as discussed in detail in chapter 4). The sixth and final positional value is 26, the *gematria* of God's essential Name, *Havayah*. The sum of the last two positional values, 23 and 26, is 49, or 7^2.

The sum of all 6 positional values is 86, the value of the Name

► We have seen that the *Shechinah*, **the Divine Presence**, is equal to 385, the square-pyramid of 10. Another way to describe a square-pyramid is as the "triangle of squares."

The Divine Presence is also known as a "bride" (כַּלָה), whose value is 55, which is the triangle of 10!

Now note that 7 · 385 = 49 · 55, as above, meaning that 385 = 7 · 55. In other words, 🔲10 = 7 · △10. A relationship between square-pyramids (or "triangles of squares") and triangles exists. Beginning with 1, every third square-pyramidal number (🔲n) is a multiple of the corresponding triangular number (△n). Thus 30 (🔲4) equals 3 · △4 and 140 (🔲7) equals 5 · △7. But only 385 (🔲10) is 7 times its corresponding triangular number, △10. Incidentally, beginning with the emotive *sefirot* (representing the 6 days of creation), kingdom (representing the Divine Presence and the "bride") is considered the seventh of the *sefirot*, just as the Sabbath is the seventh day.

---

7. *Tikun* 26 (71b).

Elokim (אֱ־לֹהִים), the Name of God that appears in the first verse (and exclusively throughout the first account of creation). The sum of the first, third, and fifth positional values is 36 (= 6²), also the value of the first three letters of *Elokim* (אֱ־לֹה). The sum of the second, fourth, and sixth values is 50 (= 2 · 5², a double-square), the value of the final two letters of *Elokim* (ים).

The series of positional values illustrates another important facet of the Name *Elokim*. The Name *Elokim* (אֱ־לֹהִים) is equal to the idiom "vessel of *Havayah*" (כְּלִי יְהוה) numerically demonstrating that the Name *Elokim* serves as a vessel for the light of *Havayah*, God's essential Name. Thus, *Elokim* refers to a more externally pronounced state of Divine manifestation in reality, expressing in a more concrete way the Divine essence referred to by the Name *Havayah*. The division of *Elokim* into the two words "vessel of *Havayah*" (כְּלִי יְהוה) is perfectly mirrored in the positional values, because the sum of the first five is 60—the value of "vessel of" (כְּלִי)▸—while the final sixth is equal to 26—the value of "*Havayah*" (יְהוה).

Earlier, we noted that each half of the verse▸▸ contains three *alef*s— three in the words, "In the beginning God created" (בְּרֵאשִׁית בָּרָא אֱ־לֹהִים) and three in the words, "the heavens and the earth" (אֵת הַשָּׁמַיִם וְאֵת הָאָרֶץ). In terms of content, the first half describes the Divine act of creation (the act of creating something from absolute nothing that only God's very essence can perform[8]). The second half depicts the totality of creation itself as consisting of two general states of reality, "the heavens and the earth," symbolizing spiritual reality and physical reality.

The positional values of the three *alef*s in the verse's latter half are 15, 23, and 26. Let's take a closer look at these numbers and the meaning they harbor. 15 is the value of God's Name *Kah* (יָה). The Name *Kah* is quite rare in the Bible, appearing in only three books: Exodus, Isaiah, and Psalms. Even more rare is the appearance of the Name *Kah* together with the Name *Havayah* (only twice in the

▸ Note also that 60 = 12 · 5, where 12 is the sum of the first two positional values, 3 and 9.

▸▸ The cantillation mark called **the etnachta** divides verses in the Bible in two. According to the rules of cantillation, the word under which this mark appears belongs to the first half of the verse.

8. See *Tanya, Igeret Hakodesh* 20.

▸ **The Name *Kah*** also appears in two more instances. The one in Jeremiah (2:31), the other in the Song of Songs (8:6), but in both cases, the grammarians (*Radak*, and others) consider it to be part of a compound word.

▸▸ The word "**worlds**" appears in the plural form and following the general rule that a plurality is interpreted in the minimal sense, this implies that God formed two worlds.

▸▸▸ The sum of the positional values, 64, is not only a square number, it can also be expressed as 4^3, a cube, and as 2^6. A square number corresponds to a two-dimensional plane, a cube to three-dimensional space, and 2^6 to a six-dimensional Hamiltonian space.

Bible[9]). From one of these, "With *Kah Havayah* formed worlds"[10] (כִּי בְּיָהּ יְהוָה צוּר עוֹלָמִים), the sages[11] learn that with the two letters of *Kah*, *yud* and *hei*, God formed two worlds. Thus, the two letters of *Kah* (יָהּ), and hence the positional value 15 in this case, allude to another duality within reality—the two worlds that God created—the World to Come (corresponding to "the heavens") and this world (corresponding to "the earth"). The third positional value, 26, is equal to the value of God's essential Name, *Havayah* (יהוה), and so 15 and 26, the first and third positional values, clearly allude to the verse, "With *Kah*◂ *Havayah* formed worlds."◂◂

Now, note that the sum of the 3 positional values of the last 3 *alef*'s is, $15 + 23 + 26 = 64 = 8^2$◂◂◂ a perfect square (indicating that they complete one another as a single unit).

Let us now see what happens when we square each positional value of the *alef*s in the verse's second half:

- $15^2 = 225$
- $23^2 = 529$
- $26^2 = 676$

The sum of these 3 squares is 1430, the product of 55 and 26, the values of "all" (הכל) and "*Havayah*" (יהוה), respectively, a product that alludes therefore to the phrase, "All is *Havayah*" (and in reverse, "*Havayah* is all"). This is one of the key phrases repeated by the followers of the Ba'al Shem Tov in the spirit of his teachings, with the purpose of strengthening our faith that there is nothing but God.

A quadratic series from positional values

Furthering our exploration of the positional values in the second half of the verse is the quadratic series we can construct from them.

---

9. Isaiah 12:2 and 26:4.
10. Ibid.
11. *Menachot* 29b.

Skipping a few stages, we here show that this series includes only 7 positive numbers,

2	15	23	26	24	17	5
13	8	3	-2	-7	-12	
-5	-5	-5	-5	-5		

The first thing we would like to note is that 26 is the zenith of this series, which allows us to correspond each of the 7 positive numbers in it to the letter in that position in the Torah's first verse. Let's highlight these letters:

בְּרֵאשִׁית בָּרָא אֱ־לֹהִים אֵת הַשָּׁמַיִם וְאֵת הָאָרֶץ

The 7 letters in these positions are, ר י א ה א ת ת, and their sum is 618, a clear reference to the Golden Ratio (recall that the Golden Number, known as *phi*, is 1.618… while its inverse is 0.618), one of the most beautiful mathematical phenomena engraved in nature.

Now let us look at these 7 positive numbers as they appear in the series. 26, the positional value of the 6th *alef* in the Torah's first verse, is situated in the middle. 26, is of course the value of *Havayah*. Incredibly, the sum of the 6 numbers around 26 is 86, once again the value of *Elokim* (אֱ־לֹהִים). What this also means is that the sum of all 7 positive numbers in this series is 112,▶ which equals the value of God's "full Name," *Havayah Elokim* (יְהוָה אֱ־לֹהִים), which first appears in the first verse of the second account of creation.

The first positive number in this series, 2, alludes to the major letter *beit* (= 2), the Torah's first letter—the first letter of "In the beginning" (בְּרֵאשִׁית). The last positive number in the series, 5, as well as the series' base: -5, both allude to the minor *hei* (= 5) of the word, "upon being created" (בְּהִבָּרְאָם) appearing in the second account's first verse,[12] a verse we will soon focus on.

Another positive number in this series is 17, the value of, "good" (טוֹב), and the square root of 289, the *gematria* of the words, "God created" (בָּרָא אֱ־לֹהִים), two of the words from the first half of the

▶ In Kabbalah, **the number 112**, the value of God's full Name, *Havayah Elokim* (יְהוָה אֱ־לֹהִים) is further related to the name Yabok (יַבֹּק), appearing later in Genesis (32:23) understood as an acronym for the words, "unification, blessing, holiness" (יִחוּד בְּרָכָה קְדֻשָּׁה).

---

12. Genesis 2:4.

Torah's first verse. 17 is also the *gematria* of the initial letters of all the words in the verse's second half, "the heavens and the earth" (אֶת הַשָּׁמַיִם וְאֵת הָאָרֶץ).

Finally, the sum of 26, with the two numbers surrounding it, 23 and 24, is 73, the *gematria* of "wisdom" (חָכְמָה), with which God created the world, as we have seen above. The sum of the other 4 positive numbers—2, 15, 17, and 5—is 39, the value of the phrase, "*Havayah* is one" (יהוה אֶחָד)!◄

A second quadratic series

Returning to the positional values of the three *alef*s in the first half of the verse, "In the beginning God created" (בְּרֵאשִׁית בָּרָא אֱ־לֹהִים), let us construct a quadratic series from them. As it turns out, this quadratic series has only 4 positive values,

$$
\begin{array}{ccccccc}
3 & & 9 & & 10 & & 6 \\
& 6 & & 1 & & -4 & \\
& & -5 & & -5 & &
\end{array}
$$

It is extraordinary that the base of the positional value series in both halves of the verse is identical: -5. There is no mathematical necessity for this to be the case.

The sum of the four positive numbers in this series is 28 (4 times 7, the average value of the 4 numbers). But, recall that the sum of the 7 positive values in the quadratic series formed by the positional values of the *alef*s in the verse's second half was 112. So the positive values in the two series have a relationship of 28:112, which is actually a ratio of 1:4. Recall that we began our study of the first verse of the Torah with the secret of the ratio 1:4, the most perfect ratio in the Torah, as it is the ratio of letters to word in the Tetragrammaton, God's essential Name, *Havayah*.

Furthermore, the sum of 28 and 112 is 140. Like 385 above, 140 is a pyramidal number, specifically,

▶ The value of the triangles of all 7 numbers equals **1218**, the lowest common multiple of 42 and 58, a most important pair of numbers in Kabbalah.

$$\text{❖}7 = 1^2 + 2^2 + 3^2 + 4^2 + 5^2 + 6^2 + 7^2 = 140$$

140 is also the union of "wisdom" (חָכְמָה) and "understanding" (בִּינָה) whose values are 73 and 67 respectively. The union of these two *sefirot* is also the union of the father and mother principles, the continual spiritual union by which the world is continuously being recreated *ex nihilo*.

Positional value relative to the words

Now let us examine the positions of the 6 *alefs* relative to the six words in which they appear.

בְּרֵאשִׁית בָּרָא אֱ־לֹהִים אֵת הַשָּׁמַיִם וְאֵת הָאָרֶץ

These word-relative positional values are then,

$$3, 3, 1, 1, 2, 2$$

Clearly revealed here is a pattern of three pairs of numbers, a beautiful phenomenon in and of itself.

The sum of these six positional values is 12, which means that their average value is 2, the value of the first letter of the Torah, the major *beit* of, "In the beginning" (בְּרֵאשִׁית).

Following the lead of our analysis regarding the verse-relative positional values, let us square these 6 numbers,

$$3^2 + 3^2 + 1^2 + 1^2 + 2^2 + 2^2 = 28$$

But, 28 is of course the total number of letters in the entire first verse. When we multiply the pairs of numbers by one another and sum, we find that,

$$3 \cdot 3 + 1 \cdot 1 + 2 \cdot 2 = 14$$

14 is the number of letters in each half of the verse.

Dividing the verse in two

Let us once again divide the Torah's first verse in two, and look at the positional values of the *alef*s relative to the 14 letters in each half. We find that these values are, 3, 9, 10 and 1, 9, and 12.

The sum of the positional values for both halves is identical,

$$3 + 9 + 10 = 22 = 1 + 9 + 12$$

The common number that appears in each half is the second number, 9, with two numbers, whose sum is 13, around it.

If instead of adding we multiply, we find that the product of the positional values of the first half of the verse is, $3 \cdot 9 \cdot 10 = 270$, or $5 \cdot 54$. The product of the positional values of the second half is $1 \cdot 9 \cdot 12 = 108$, or $2 \cdot 54$. The sum of the two products is then 378 the *gematria* of the word "electrum" (חַשְׁמַל)◄, the most mysterious word in the Bible that appears three times in the Working of the Divine Chariot. 378 is also the triangle of 27.

378 has a special connection with the letter *alef* (א). The value of the letter *alef* together with its first and second filling is 378!

א + אלף + אלף למד פא = 378◄◄

Returning to the six positional values relative to the two halves of the verse, we find that their sum is 44. When we add this to the sum of the positional values relative to the entire verse, 86, the total is $130 = 10 \cdot 13 = 5 \cdot 26$. When added to the sum of the positional values relative to the words in the verse, 12, the total is 56, the diamond form of 7.

► $378 = 7 \cdot 54$, with $2 \cdot 54$ in the first half and $5 \cdot 54$ in the second. This division of 7 into 5 and 2 is the secret of the word "**gold**" (זָהָב), since the values of its three letters are 7, 5, and 2, alluding to the mathematical identity: $7 = 5 + 2$.

►► The value of 6 letters *alef* calculated in this way is then, $6 \cdot 378 = 2268 = 7 \cdot 324$. 324 is the square of 18, the value of "life" (חַי).

THE FIRES OF CREATION

Letter permutation

The year was 1716 and Rabbi Yisrael Ba'al Shem Tov who had just turned 18 years old, was chosen to lead the group of hidden *tzadikim* (righteous and holy individuals), the members of a secret brotherhood. Members of the secret brotherhood usually lived alone in the forests of Eastern Europe, preferring the solitude that would allow them to serve God and learn Torah quietly over the hustle and bustle of the growing urban centers where Jewish communities were traditionally located. The young leader's first suggestion was that in the interest of spreading love of God and Torah to the common folk, the *tzadikim* would change their secluded lifestyle and move to the cities and towns that had Jewish communities. Being close to their Jewish brethren would allow them to extend their spiritual influence and blessings to all Jews.

How could the members of the hidden *tzadikim* exchange the quiet solitude of their forest-life with the socially taxing life of growing communities? One might expect the Ba'al Shem Tov's "experiment" to end in failure. But, the opposite was true. The hidden *tzadikim* followed his suggestion and changed Jewish life forever by forming the backbone of the Chassidic movement. How could the Ba'al Shem Tov have known that such a change in lifestyle and purpose would not leave the hidden *tzadikim* feeling detached and lost?

We do not know the exact answer to this question, and certainly the Ba'al Shem Tov's *ru'ach hakodesh* (holy spirit of understanding) played an essential part, we can speculate that his suggestion was also based on a simple letter permutation. Whereas the word "forest" is spelled יַעַר, permuting the first two letters yields the word "city" spelled עִיר.▸ This simple letter permutation illustrates an

▸ Apart from promoting the movement of the hidden *tzadikim* to the city, **the Ba'al Shem Tov** has another important connection to the Hebrew word for "city" (עִיר). The Ba'al Shem Tov was born in the year 5458 (1698), the *gematria* of the word נְחַת (in Aramaic), which literally means, "to descend." In the Book of Daniel we find this word in the verse, "A holy angel descended from heaven" (עִיר וְקַדִּישׁ מִן שְׁמַיָּא נָחִת). This verse is seen as an allusion to the Ba'al Shem Tov (likened to the angel descending from heaven) in the year 5458, נְחַת. Of course, the word for "angel" (עִיר) is the same as the word for "city" (עִיר).

essential ingredient in the Ba'al Shem Tov's spiritual path. As much as spiritual ascent is dependent on the *tzadik's* ability to divorce himself from the time-consuming and confusing demands of society and livelihood, the ascent is not complete if it does not lead the *tzadik* to descend back into his community, illuminating the lives of his fellow Jews with his sincere dedication to serving God, loving his neighbors, and learning Torah. The forest (יַעַר), in this case, signifies the tzadik's spiritual ascent, while the city (עִיר) signifies his descent back into mundane reality, bringing with him the treasures acquired above. It also illustrates how analyzing situations and events in Hebrew allows their essence to be identified and even transmuted into a similar, related essence.

Covenant of fire

Letter permutation is one of the most central forms of textual analysis in the *Zohar* and its accompanying volumes. The *Tikunei Zohar* bases its 70 interpretations of the Torah's first word on many of its possible letter permutations.

Since the Torah's first word, "In the beginning" (בְּרֵאשִׁית) has 6 (different) letters, there are a total of 6! = 720 possible permutations. One of these permutations[1] is formed by taking the first and final two letters of the word בְּרֵאשִׁית and combining them into one word, with the two middle letters forming a second word.

Let us picture this process in three steps:

בְּרֵאשִׁית

בר אש ית

בְּרִית אֵשׁ

The resulting two word idiom בְּרִית אֵשׁ means "covenant (בְּרִית) of fire (אֵשׁ)." And so we learn that the world was created with a covenant of fire.

---

1. Introduction to *Tikunei Zohar* (10b).

▶ The second account of creation describes Adam's creation with the following words (Genesis 2:7), "And *Havayah Elokim* formed Adam dust from the earth…." Unlike its appearance everywhere else in the Bible, the verb "formed" (וַיִּיצֶר) in this verse is written using two *yuds*. This unique form of the verb containing two *yuds* is explained by the sages as alluding to two formations, "formation in this world and formation for the World to Come" (*Bereisheet Rabbah* 14:5). Thus, when God formed Adam, dust from the earth, He already prepared man for **two lifetimes**, one in this world and the other in the World to Come.

When the Name of the Divine Presence, *Kah* (יה), resides between them, a man and woman can also bestow upon their offspring two formations, one for this world and one for the World to Come. The formation in this world derives primarily from the *hei* of "woman" (אִשָּׁה) and the formation for the World to Come derives primarily from the *yud* of "man" (אִישׁ).

The fiery covenant of matrimony

The covenant of fire in its most pristine form is associated with the covenant that binds two people together in matrimony. In chapter 7, we saw how "the way of a man with a maiden" corresponds to the element of fire and alludes to God Himself who is likened, as it were, to fire.

How do we see that it is fire that binds a married couple together? In Hebrew, the two consonant letters used to spell both "man" (אִישׁ) and "woman" (אִשָּׁה) are *alef* (א) and *shin* (שׁ), which together spell "fire" (אֵשׁ). The additional vowel letter in "man" is the letter *yud* (י) and the additional vowel letter in "woman" is the letter *hei* (ה), which together form God's Name *Kah* (יה), the Name with which God formed two worlds, the World to Come (with the *yud*) and this world (with the *hei*), as we saw above.▲

Based upon this phenomenon the sages teach that,[2] "When they [the man and the woman] merit, the Divine Presence resides between them [the Name *Kah*], but when not, fire consumes them." So, though fire is always at play in the relationship of a married couple, when their bond is based on serving God and leading an upright life, the innate fire is sanctified and becomes a source for elevating them in their pursuit of Godliness; like the fire that consumed and elevated the sacrifices placed on the altar in the Temple. Their two fires unite in the secret of the "covenant of fire" (בְּרִית אֵשׁ), God's covenant with the world created▶▶ for His glory (a connotation of the

▶▶ Two of the 22 letters of the *alefbeit* possess a most beautiful **symmetric form**. These are the letters *alef* (א) and *shin* (שׁ), which together spell "fire" (אֵשׁ), the covenant of creation! A couple decides to marry, to create a covenant when the one finds favor in the other's eyes; when the one reflects the other's inner being—in short, when they are symmetric.

In Kabbalah, the symmetry of the *alef* is the "symmetry" exhibited by God's infinite light before the contraction (צִמְצוּם) of the light that began creation (the *alef* is before the Torah's opening letter, the *beit* of "In the beginning" [בְּרֵאשִׁית]).

The symmetry of the *shin* is the symmetry within creation, after the contraction of the infinite light. The first, we will refer to as *infinite* symmetry and the second, as *finite* symmetry. Only when the infinite symmetry is broken as it were in the secret of the initial contraction, beginning with a shift in the infinite light from symmetry to asymmetry, can finite symmetry appear. This secret is alluded to in the Torah's first word, "In the beginning" (בְּרֵאשִׁית), whose letters can also be permuted to spell the two phrases: "I broke [the] *alef*" (שָׁבַרְתִּי א), and "I created [the] *shin*" (בָּרֵאתִי ש)!

2. *Sotah* 17a.

▶▶ 6 times "fire" (אֵש), 301 equals 1806 = 21 · 86, where 21 is God's Name *Ekyeh* (אֶהְיֶה), the Name associated with the redemption and 86 is the value of His Name, *Elokim* (אֱ־לֹהִים), the Name associated with creation.

Divine Presence, the Divine "bride").[3] For this reason the world was created with a *beit*—the first letter of "In the beginning" (בְּרֵאשִׁית)—whose numerical value, 2, alludes to the union of male and female throughout creation.

Six fires

In the tractate of *Yoma*,▼ the sages describe six different types of fire.[4]◀◀ Each of these six fires reveals an aspect of the covenant of fire with which God created the world.

In the remainder of this chapter, we will contemplate six pairs of related words or proper nouns whose initial letters spell the word "fire" (אֵש). Each pair of words alludes to one of six types of fire and thus to the six aspects of the covenant of fire alluded to in the Torah's first word, which can be read as, "He created six" (בָּרָא שִׁית).

The first covenant of fire: *Kel Shakai*

The first pair of words whose initials spell "fire" (אֵש) that we will contemplate are two of God's Names—אֵ־ל (pronounced *Kel*) and שַׁ־דָּי

3. Isaiah 43:7. See also *Avot* 6:11.
4. *Yoma* 21b. See in length in our Hebrew volume, *Sod Hashem Liyerei'av*, pp. 245ff.

▶ **The tractate of *Yoma***, which literally means, "the day" is so named in reference to the holiest day of the year, Yom Kippur, the topic discussed throughout most of it.

On folio 21b, the sages discuss various qualities of the eternal fire burning on the Temple's outdoor altar, the altar upon which the various sacrifices were placed in order to be consumed. From that discussion they move on to describe six types of fire and these are their words:

Our masters taught: There are six [types of fire]: (1) There is that which eats but does not drink; (2) there is that which drinks, but does not eat; (3) there is that which eats and drinks; (4) there is that which eats both dry and damp; (5) there is fire that repels fire; (6) there is fire that eats fire. Fire that eats but does not drink is like our [physical] fire. Fire that drinks but does not eat is the fever of the sick, who desire to drink but not to eat. That which

eats and drinks is Elijah's fire, as the verse says, "And the water that was in the canal, the fire ate." That which eats both dry and damp is the fire on the altar. The fire that repels fire is the fire of the angel Gabriel, who upon entering the furnace into which the three tzadikim Hannania, Mishael, and Azaria had been cast repelled the fire from inside the furnace to the outside to burn those who had thrown them in. The fire that eats fire is the fire of the Shechinah [Divine Presence], as was said, "He put His finger between them [the angels who opposed the creation of man; angels are likened to fire] and burnt them."

The numerical value of the phrase, "there are six fires" (שֵׁש אֵשׁוֹת הֵן) is 1362 = 6 · 227, the *gematria* of the word "blessing" (בְּרָכָה), revealing that each of these holy fires is a source of blessing. Indeed, the sages teach that the *beit* (ב), the Torah's very first letter, stands for "blessing" (בְּרָכָה).

(pronounced *Shakai*). These two Names appear together as a composite Name 7 times in the Bible[5] (6 in the Pentateuch and 1 in the Prophets).[6] The 6 appearances of *Kel Shakai*▼ in the Pentateuch may be seen to correspond to all 6 pairs of words whose initials spell "fire," i.e., to the 6 fires of creation, and in this sense this is the all-inclusive covenant of fire.

One of the principles set forth by the sages in analyzing the Torah is that the first time that a word or a phrase appears in the Torah determines its essence and guides our understanding of its subsequent appearances.[7] The first appearance of *Kel Shakai* in the Torah is in the verse,

> *Havayah appeared to Avram and said to him, "I am Kel Shakai, walk before Me and be complete."*[8]

These words begin the covenant of circumcision God made with Abraham, the first Jew. At this juncture, his name was changed from Avram to Avraham (pronounced, Abraham, in English

---

5. Genesis 17:1, Ibid. 28:3, Ibid. 35:11, Ibid. 43:14, Ibid. 48:3, Exodus 6:3, and Ezekiel 10:5.
6. "All sevens are endeared" (*Vayikra Rabbah* 29:11).
7. In the words of the sages, "all follows the root" (הַכֹּל הוֹלֵךְ אַחַר הָעִקָּר) from *Ma'asrot* 3:10 and "the body follows the head" (בָּתַר רֵישָׁא גּוּפָא אָזִיל) from *Tosefta Ta'anit* 2:6.
8. Genesis 17:1.

▶ When each of the letters of *Kel Shakai* (אֵיל שַׁדַּי) is written out, the complete filling contains 15 letters: אלף למד שין דלת יוד. 15 letters can be arranged in the form of a triangle, **the triangle of Kel Shakai**, like so,

$$
\begin{array}{ccccccc}
 & & & א & & & \\
 & & ף & & ל & & \\
 & ד & & מ & & ל & \\
 ש & & י & & ן & & י & & ד \\
ל & & ת & & י & & ו & & ד
\end{array}
$$

First note that structurally, arranging the letters in this form divides the two Names neatly between the rows with the filling of *Kel* (אלף למד) populating the first three lines of the triangle, and the filling of the Name *Shakai* (שין דלת יוד) populating the last two.

The letters in the vertices spell the mystical Name *Ailaid* (אלד), which means "I will give birth"; the three letters in the middle of each of its three sides spell "child" (יֶלֶד); the three letters down the middle axis of the triangle spell "my mother" (אִמִּי). Together, these three words read, "my mother [says] I will give birth to a child" capturing in a nutshell the intention and purpose of the marital covenant alluded to by the Name *Kel Shakai*.

Other notable phenomena in this triangle are that the value of the three middle letters (מין) is 100, or 10^2. The value of the entire triangle is 999 (27 times 37; 37 is the value of Abel, the son of Adam and Eve, as we have discussed above) or one less than 10^3. Also, in another allusion to the covenant this Name is associated with, the three corner triangles of 3 letters each (**אלף שלת דוד**) equal 855 = 45 · 19, or the product of "Adam" (אָדָם), 45 and "Eve" (חַוָּה), 19. The remaining 6 letters (**למדין**), form an upside-down triangle, and their sum is 144 = 12^2.

▶ The *gematria* of Abraham (אַבְרָהָם), 248, is the same as that of "**in God's image**" (בְּצֶלֶם אֱ-לֹהִים), the exact wording used by the Torah to describe the Divine image with which God created Adam and which gave him a special and unique role in creation. Though Adam lost the manifestation of the Divine image through his sin, Abraham, through his pure faith and willingness to follow God's will unquestionably, became its essential propogator.

▶▶ The first three letters of **Abraham** (אַבְרָהָם) permute to spell the root "to create" (בְּרָא). These are also the first three letters of the Torah (בְּרֵאשִׁית) and the complete second word of the Torah, "created" (בְּרָא). In the order they appear in "Abraham," these 3 letters spell the word "organ" (אֵבָר), which specifically refers to the organ of procreation, the organ of circumcision.

The remaining two letters of Abraham (אַבְרָהָם) equal 45, the *gematria* of Adam (אָדָם), i.e., "man." Thus, Abraham is considered the (all-inclusive) organ of mankind, alluding to the organ of circumcision.

The *gematria* of Abraham's entire name is 248, the total number of organs in the human body created "in the image of God" (בְּצֶלֶם אֱ-לֹהִים), whose value is also 248!

transliteration).[9] It was here that "the image of God,"[10] in which God created man was revealed in him, and here he became potent, able to procreate, "in God's image."[11] ◀

The word "covenant" (בְּרִית) is commonly associated with circumcision even though this is not its first appearance in the Torah, because it appears in the context of God's command that Abraham circumcise himself and his household with such high frequency (13 times). Since the covenant of fire that is *Kel Shakai* can be soundly identified as the covenant associated with creation, the Torah's first word, "In the beginning" (בְּרֵאשִׁית) can now be homiletically associated with the covenant of circumcision. Since Abraham was the first human commanded to circumcise and God made the covenant of circumcision directly with him, this deepens our awareness of Abraham's soul as instrumental in the process of creation. ◀◀

As it is, Abraham's name and essence are strongly identified with creation. The second account of creation begins with the words,

These are the generations of the heavens and the earth upon being created, on the day that Havayah Elokim made earth and heavens.

The sages note that the letters in the word meaning "upon being created" (בְּהִבָּרְאָם) permute to spell "with Abraham" (בְּאַבְרָהָם),[12] from which we learn that the world was created with Abraham's essence: loving-kindness,▼▼▼ considered the building-block of reality as revealed in the verse, "The world shall be built by loving-kindness."[13]

9. Ibid. 17:5.
10. Ibid. 1:27 and 9:6.
11. Ibid. 5:1. In this context, the first ten generations of man are enumerated beginning with Adam who is described as having been created, "in the likeness of God."
12. *Zohar* I, 3b.
13. Psalms 89:3.

▶▶▶ The essential character trait of each of the Patriarchs corresponds with one of the emotive *sefirot*. **Abraham's essential loving-kindness** is learnt from the phrase, "Loving-kindness to Abraham" (Micah 7:20). His ability to shower loving-kindness upon complete strangers served him in his goal of spreading pure monotheistic faith in God (see Genesis 21:33 and *Rashi* there). Even on the third day following his circumcision, Abraham sat anxiously at his tent entrance, waiting for the opportunity to host and care for any desert wanderers that might pass by.

The first five appearances of the Name *Kel Shakai* are all in Genesis, in the stories of the Patriarchs. The sixth and final appearance is found at the beginning of the second Torah portion of Exodus, in the verse,[14]

> *And I appeared to Abraham to Isaac and to Jacob as Kel Shakai,*
> *but by My Name Havayah, I was not known to them.*

The *gematria* of *Kel Shakai* (אֵ־ל שַׁ־דַּי) is 345 = Moses (מֹשֶׁה).▸

Just as the world was created with Abraham's soul-root, so the Divine revelation to the Patriarchs, Abraham, Isaac, and Jacob, was through the conduit of Moses' soul-root, the redeemer of Israel (to whom was revealed the higher Name *Havayah*, not revealed to the Patriarchs).

In Kabbalah, it is explained that the Name *Kel Shakai*▸▸ originates in understanding (as it is said, "And the soul of *Shakai* gives them understanding"▾▾▾ [וְנִשְׁמַת שַׁ־דַּי תְּבִינֵם]), the "concealed world" (עָלְמָא דְּאִתְכַּסְיָא),[15] the level of consciousness that Moses merited attaining, or in the commonly used idiom, "Moses merited understanding"▸▸▸▸ (מֹשֶׁה זָכָה לְבִינָה).[16]

In the first account of creation, which spans the first chapter of

---

14. Exodus 6:3.
15. See *Etz Chaim* 3:2.
16. *Etz Chaim*, Klalim and 15:6. See *Zohar* II, 23b.

▸▸▸ The verse, "Verily, spirit is in man and the soul of *Shakai* gives them understanding" comes from the Book of **Job** (32:8), which relates how after being stricken with terrible affliction, Job's three best-friends come to be with him. For a few days Job remains silent, but when he finally begins to talk, he reveals his bitterness and anger at God. Job's friends try to rationalize his suffering, but fail.

Sitting aside and quietly listening to the dialogue between Job and his friends was Elihu ben Berachel. Finally, Elihu interjects and apologizes for his words, stating that though he hoped that wisdom would be with his elders, they have failed in responding properly to Job. The *gematria* of Elihu's full name (אֱלִיהוּא בֶּן בְּרַכְאֵל) is 358, the same as Mashiach (מָשִׁיחַ), alluding to Mashiach's ability to comfort us after all the suffering of our exile. Elihu opens with the words, "Verily, spirit is in man and the soul of *Shakai* gives them understanding."

▸ The *gematria* of *Kel Shakai* when written in full (אלף למד שין דלת יוד) is 999. The Arizal explains that 999 plus 1 (the *kollel*, representing the Name itself) alludes to **the 1000 lights** that were given to Moses at Sinai.

▸▸ The two parts of the compound Name, **Kel Shakai** suggest that *Kel* (אֵ־ל), the Name associated with the *sefirah* of loving-kindness, flows into and unites with the *Shakai* (שַׁ־דַּי), the Name associated with the *sefirah* of foundation.

▸▸▸▸ The value of the complete phrase, **"Moses merited understanding"** (מֹשֶׁה זָכָה לְבִינָה) is 474, also the value of "knowledge" (דַּעַת), the third intellectual power. Just the first two words, "Moses merited" (מֹשֶׁה זָכָה) equal 377, the value of "Samuel" (שְׁמוּאֵל), the only other prophet whose level of prophecy, while not equivalent, in a certain sense is considered similar to Moses'. The *gematria* of the third word, "[to] understanding" (לְבִינָה) is 97 (the 26th prime number, starting from 1), which is equal to the archetypal feminine figure in Kabbalah whose origin is in understanding, the mother principle, "*Meheitavel*" (מְהֵיטַבְאֵל).

The number 97 also figures prominently in this phrase in other ways: Performing a "dot-product" on the letters of the first two words, yields $40 \cdot 7 + 300 \cdot 20 + 5 \cdot 5 = 6305$, or the product of 65 and 97 where 65 is the value of God's Name, Adni (אֲדֹנָי) and 97 is once again "*Meheitavel*." In addition the "primordial value" (מִסְפָּר קַדְמִי) of the full expression is 1746, or 18 times 97.

▶ God's Name *Elokim* appears exactly 32 times in the six days of creation. These 32 appearances of *Elokim* correspond to the **32 pathways of wisdom** with which the world was created, as stated in the beginning of *Sefer Yetzirah*, whose contents are attributed to Abraham.

▶▶ "God is good to all and His mercy is upon all He has made."

טוב יהוה לכל ורחמיו על כל מעשיו

The *gematria* of the complete verse is 969 = 57 · 17, where 17 is the value of the first word in the verse, "good" (טוב).

But, 969 is not just a multiple of 17, i.e., "good," it is actually **the tetrahedron** of 17, meaning that it is the sum of all triangles from the triangle of 1 to the triangle of 17. Using our notation for figurate numbers, we can define the tetrahedron of n as:

$$\triangle n = \sum_{k=1}^{n} \triangle k$$

Genesis and the first 3 verses of chapter 2, only God's Name *Elokim* appears. ◀

The *gematria* of *Elokim* (אֱ־לֹהִים) is 86, the same as the Hebrew word for "nature" (הַטֶּבַע).[17] In the first verse of the second account of creation (which is also considered the concluding verse of the first account, indicating that in essence it is the verse that connects the two accounts together, resolving the apparent differences between them), where God's Name *Havayah* appears for the first time in the Torah, the word "upon being created" alludes to Abraham's soul-root, as mentioned above. The Name *Elokim* corresponds to the Divine attribute of judgment and law (in nature, it represents the laws of nature), whereas the soul-root of Abraham is the Divine attribute of loving-kindness. Through Abraham it becomes manifest that love—i.e., God's love for His world—is what actually motivates judgment and law in nature. Yet, until the appearance of Abraham on the stage of history, 20 generations after the world's creation, God's essential loving-kindness for all of His creations remained concealed from human consciousness. Abraham was the first to reveal to mankind that God is omnipresent and that His Providence looks over and, with infinite love, cares for all, "God is good to all, and His mercy is upon all He has made."[18] ◀◀

Everything that happens in the natural order is in truth, in depth, a miracle enclothed in the garb of nature so as not to be apparent to everyone. This is the meaning of the Name *Kel Shakai* that was revealed to the Patriarchs—"concealed miracles (*Shakai*) from love (*Kel*)." But the Name revealed to Moses (and the People of Israel) at the time of the Exodus, the higher Name *Havayah*, is the Name that performs miracles that entirely transcend and defy the laws of nature. This Name was the source of the Ten Plagues wrought on Egypt and the Splitting of the Red Sea, both sets of miracles by definition were meant to defy nature, in order to reveal that God is

17. *Megaleh Amukot parashat Chukat.*
18. Psalms 145:9.

above nature, He creates it and He can put its laws on-hold, so to speak.

Thus, the first "covenant of fire" alluded to in the first word of the Torah, the beginning of creation, is God's covenant with His world, as expressed by miracles enclothed in the garb of nature and nature's laws (beginning with the continuous miracle of creation itself), deriving from God's infinite, fiery love for all of His creation:

בְּרִית אֵ־ל שַׁ־דָּי

The second covenant of fire: Something and Nothing

The world was created *ex nihilo*, "something from nothing" (יֵשׁ מֵאַיִן). The word for "nothing" (אַיִן) begins with an *alef*. Although the letter *alef* is normally a vowel letter, here it is the more emphasized consonant of the word, as evident from the fact that even when the word's final *nun* is dropped▶, the resulting word, אִי still means "nothing."

"Something" (יֵשׁ) ends with a *shin* (שׁ). In *Sefer Yetzirah*, we learn that the letter with which God creates fire is the letter *shin*, the final letter of the word "fire" (אֵשׁ) and the more pronounced consonant in this word (though the initial letter, the *alef*, also serves as a consonant, it is less pronounced than the *shin*). And so with regard to "something" (יֵשׁ), the more pronounced consonant is the *shin*, the final letter of the word, not the *yud* (which usually serves as a vowel; note also that *alef* and *yud* interchange in Hebrew).

Thus the pair of words, "nothing" (אַיִן) and "something" (יֵשׁ), which hold within them the secret of creation "something from nothing" (יֵשׁ מֵאַיִן)▶▶–"nothing" beginning with an *alef* and "something" ending with a *shin*—alludes to "fire" (אֵשׁ). Our second covenant of fire is thus, "a covenant of something from nothing":

בְּרִית אַיִן יֵשׁ

Note that the word for "man" (אִישׁ), which in Hebrew means not

▶ Dropping **the letter *nun*** (נ) is not arbitrary, because this letter is associated with falling or dropping. *Nun* is the first letter in "fall" (נָפַל). The sages (Berachot 4b) ask why chapter 146 of Psalms—which is written in acrostic form—lacks a verse beginning with the letter *nun*. They reply that the letter *nun* is associated with falling as learned from the verse, "Fallen, she shall not rise again, the maiden of Zion" (Amos 5:2).

Indeed, in Hebrew grammar, the *nun* is a letter that often falls from words, i.e., disappears, without affecting the meaning of the word. One of the most interesting examples of this is in the word "truth" (אֱמֶת). According to Rabbi David Kimchi, the *Radak*, one of the greatest grammarians of the Middle Ages, the word "truth" stems from the root meaning to believe (א.מ.נ.) appearing in the form אמנת, but with the *nun* falling off.

▶▶ The idiom "**something from nothing**" (יֵשׁ מֵאַיִן) equals 411, also the *gematria* of the Torah's tenth word "chaos" (תהו), pronounced *tohu*, in Hebrew. *Tohu* is the initial state of creation, before reality settles down and assumes fixed form (similar to how scientists today describe the first moments after the Big Bang). *Tohu* actually precedes light, and is referred to as the "darkness" that precedes light (see for example Isaiah 45:7). The average value of the three letters of "chaos" (תהו) is 137, the inverse of the fine structure constant, arguably the most important pure number in creation.

▶ The 6 letters of the two words "**individual**" (אִישׁ) **and** "**species**" (מִין) can be permuted to spell "something from nothing" (יֵשׁ מֵאַיִן). Each individual part of creation reflects the mystery of the Workings of Creation, creation *ex nihilo*, with the general, abstract category of the species implicit in the reality of each individual specimen.

▶▶ Fire is a symbol for energy. Energy appears in two basic forms: **potential and kinetic**. Relative to one another, potential energy is "nothing" (in both the objective and subjective senses of the word, as discussed in the text) while kinetic energy is "something." The *alef* of "fire" (אֵשׁ) represents potential energy while the *shin* of "fire" (אֵשׁ) represents kinetic energy.

▶▶▶ The sudden appearance of something within the void is similar to the event described in modern physics as the spontaneous emergence of an elementary particle out of empty space.

only a human male but a specimen (an individual member of a "species"[19]◀), a concatenation of the words "nothing" (אַיִן) and "something" (יֵשׁ), as though to say that every specific aspect of reality is continuously being created "something from nothing." Like before, the *nun* of "nothing" (אַיִן) falls out of the word and the *yud* in "nothing" (אַיִן) joins the *yud* in "something" (יֵשׁ) to become one in the *yud* in "man" (אִישׁ).

In the idiom, "something from nothing," "nothing" carries two meanings, one objective the other subjective.◀◀ The objective meaning of "nothing" is that indeed all of reality was created (and is continuously being recreated) from absolute nothingness. The subjective meaning occurs from the perspective of the created being, whose consciousness remains oblivious to its source in Divinity, i.e., God's infinite light. Because the Creator and His light—the source from whence all comes into existence—remain hidden from creation, from this perspective, all was created out of a subjective sense of nothingness. God allows His light to be concealed in this manner in order to grant creation an independent state of self-consciousness, and give man free choice.

In Kabbalah, we learn that the process of creation begins with the initial contraction of God's infinite light, thereby creating an empty void in which all future creation can evolve. First God created the nothing, i.e., the void, and then He created the something. This is one of the deep meanings in Kabbalah of creation having been "something from nothing." In the void, the primordial nothing, the future-something was definitely absent, but suddenly it appeared there, from nothing.◀◀◀

When the Greek king, Ptolemy, had 70 sages independently translate the Bible into Greek (a translation known as the Septuagint), all of them, by Divine inspiration, changed the order of the words in the first verse, translating it as, "God created in the beginning" (אֱלֹהִים

---

19. For a more in-depth treatment of Torah taxonomy, see the article "Torah and taxonomy" on our website.

בָּרָא בְּרֵאשִׁית).[20] Their intent was to prevent the non-Jewish (and, at the time, pagan) reader from mistakenly concluding that the words, "In the beginning God created" imply that something called the "In the beginning" created God.

Given that this translation was Divinely-inspired, in order to bring the wisdom of the Torah to the nations of the world, it merits consideration. Interestingly, it's ordering of the initial words of the Torah has the Torah begin with the first letter of the alphabet, the letter *alef* of God (אֱ-לֹהִים), instead of with the second letter of the alphabet, the *beit* of "In the beginning" (בְּרֵאשִׁית). The first two words are, "God created" (אֱ-לֹהִים בָּרָא), whose *gematria* is 289, or 17^2, where 17 is the value of "good" (טוֹב).

The first five letters of the Torah's first word "in the beginning" (בְּרֵאשִׁית), which now, in the Septuagint has become the third word, can be read (by reversing the order of the 4th and 5th letters) as "created something" (בָּרָא יֵשׁ). The second word, "created" (בָּרָא), when read separately from the third word, can be understood as implying "created nothing" (בָּרָא אַיִן). The first three words of the Septuagint variant thus allude to: "God created nothing, the secret of the void (the origin, in Kabbalah, of non-Jewish consciousness, for which the sages translated the Torah into Greek), then He created something." ▶

The *gematria* of these two phrases, "created nothing, created something" (בָּרָא אַיִן בָּרָא יֵשׁ) is 777 = 21 · 37. When we add "God" (אֱ-לֹהִים) to each phrase giving us, "God created nothing, God created something" (אֱ-לֹהִים בָּרָא אַיִן אֱ-לֹהִים בָּרָא יֵשׁ)▼▼, the value of these two

---
20. *Megilah* 9a.

▶ Adding the *gematria* of "nothing" (אַיִן) to that of "something" (יֵשׁ) gives 371, which is also 7 times the value of "stone" (אֶבֶן), alluding to (Genesis 49:24), "the stone of Israel" (אֶבֶן יִשְׂרָאֵל). In the phrase "**created nothing, created something**," the value of the two words, "created" (בָּרָא) is 406, or 7 times "grace" (חֵן), also the triangle of 28.

The *gematria* of both phrases together, "created nothing, created something" (בָּרָא אַיִן, בָּרָא יֵשׁ) is 777 and clearly it follows that 777 is also 7 times each of the words, "stone" (אֶבֶן) and "grace" (חֵן) together, or 7 times 111. When these two words, "stone" and "grace" are joined, they form the idiomatic expression for a "precious stone" (like a diamond, ruby, etc.)

In *Sefer Yetzirah* the word "stone" refers to a letter of the Hebrew alphabet, and the "precious stone" (the corner stone upon which the alphabet is based) is the letter *alef*, which when written in full, אֶלֶף, equals 111, the value of "precious stone" (אֶבֶן חֵן).

▶▶ The two expressions "**God created nothing, God created something**" (אֱ-לֹהִים בָּרָא אַיִן אֱ-לֹהִים בָּרָא יֵשׁ) comprise 21 letters, allowing us to draw them in the form of the triangle of 6

```
            א
          ל  ה
        י  ם  ב
      ר  א  א  י
    ו  א  ל  ה  י
  מ  ב  ר  א  י  ש
```

The value of the 6 letters inside the triangle (the inner triangle of 3) is 78 = 6 · 13 (the average value of each of the inner letters). The 15 letters that make up the triangle's exterior shell equal 871 = 13 · 67, where 67 is the value of "understanding" (בִּינָה), the companion of wisdom! (And, the reduced value of "understanding" is 13.)

▷▷▷

137

▶ The sum of 21 and 37 is 58, the value of "grace" (חֵן), or "Noah" (נֹחַ), whose name means "rest" (נָח). The sum of 13 and 73 is 86, the value of *Elokim* (אֱ־לֹהִים).

Together, 58 and 86 equal 144 = 12², which is the total number of letters in the 3 verses that make up the account of the seventh day of creation, the Shabbat. In these verses there is no explicit mention of the seventh day's name being Shabbat (as a noun), however, the three letters of Shabbat appear in verb form. Beautifully, the number of letters up to and including the three letters of Shabbat (שבת) in the verb "he rested" (וַיִּשְׁבֹּת) is 58, meaning that there are 86 remaining letters.

וַיְכֻלּוּ הַשָּׁמַיִם וְהָאָרֶץ וְכָל צְבָאָם. וַיְכַל אֱ־לֹהִים בַּיּוֹם הַשְּׁבִיעִי מְלַאכְתּוֹ אֲשֶׁר עָשָׂה וַיִּשְׁבֹּת בַּיּוֹם הַשְּׁבִיעִי מִכָּל מְלַאכְתּוֹ אֲשֶׁר עָשָׂה. וַיְבָרֶךְ אֱ־לֹהִים אֶת יוֹם הַשְּׁבִיעִי וַיְקַדֵּשׁ אֹתוֹ כִּי בוֹ שָׁבַת מִכָּל מְלַאכְתּוֹ אֲשֶׁר בָּרָא אֱ־לֹהִים לַעֲשׂוֹת:

Thus, the letters of Shabbat divide the 3 verses into 58 and 86 letters!

phrases becomes $949 = 13 \cdot 73$. These two pairs of numbers, 21 and 37 and 13 and 73 are particularly related. 13 and 21◄ are two consecutive numbers in the Love (Fibonacci) series of numbers; they are also two consecutive covenant numbers.

We have already discussed in length the special mathematical relationship between 37 and 73, let us recall though that their product is 2701, the value of the Torah's first verse.

The *gematria* of the expression "God created something from nothing" (אֱ־לֹהִים בָּרָא יֵשׁ מֵאַיִן) is 700, the value of "Seth" (שֵׁת), Adam's third son whom we met above in our study of mankind's first family. We will meet Seth again presently in the context of the fifth covenant of fire.

To conclude, in the soul the covenant of nothing and something is the power to actualize potential, to manifest talents, to realize desires, and to achieve fulfillment. This then is the second covenant of fire:

בְּרִית אַיִן יֵשׁ

The third covenant of fire: Earth and Heaven

The third pair of words related to fire that we will contemplate is explicit in the Torah's first verse, "In the beginning God created the heavens and the earth." They are the two initial, all-inclusive

▷▷▷ The letters in the triangle's 3 vertices are א, מ, ש, the three *mother letters* (אִמּוֹת) of the Hebrew *alefbeit*, as explained in *Sefer Yetzirah*. *Alef* stands for "air" (אֲוִיר), *mem* for "water" (מַיִם), and *shin* for "fire" (אֵשׁ), the three primary elements of creation.

The same letters can be drawn in the form of a descending triangle, whose first two lines contain the first expression and remaining four lines, the second expression:

```
א ל ה י ם ב
ר א א י
א ל ה י
ם ב ר
א י
ש
```

In this triangle, the value of the 6 inner letters is $49 = 7^2$. The value of the 15 exterior letters is $900 = 30^2$. When added, the two roots, 7 and 30, equal 37, the fundamental number of creation!

▶ The Torah begins with the verse, "In the beginning, God created the heavens and the earth." There is no mention of "light." Normally, light is associated with spirituality and the heavens more than it is associated with materiality and the earth. For this reason, when we imagine the soul we think of the heavens and light, whereas we tend to associate the body with the earth and darkness, as seen in the Torah's second verse, "And the earth was chaotic and void, with darkness upon the abyss...."

Though this association certainly has its merits, **a higher form of light is destined to emerge from earthly materiality**, specifically from the body. In the Torah, this can be seen in the third verse, which reads, "God said, 'Let there be light,' and there was light." God's words, "Let there be light," imply light originating in the heavens (from above), or what is known as direct light (אוֹר

יָשָׁר) in Kabbalah, while the words, "and there was light," imply light originating from the earth (below), or reflected light (אוֹר חוֹזֵר), in Kabbalah. In fact, the value of these words, "and there was light" (וַיְהִי אוֹר) equals Rachel (רָחֵל), the archetypal soul of kingdom and earth. Likewise, the word "earth" (אֶרֶץ) itself is known to be an acronym for the phrase, "light is sown for the tzadik" (אוֹר זָרַע לַצַּדִּיק).

Another way to see this is to look at the values of these words. The Hebrew words for "soul" (נְשָׁמָה) and "the heavens" (הַשָּׁמַיִם) have the exact same value: 395.

Though "the earth" (הָאָרֶץ) equals 296 while "body" (גּוּף) is only equal to 89, when we add "light" (אוֹר), which equals 207 to "body" (גּוּף), we do get 296. So "the earth" is equal to "light" and "body" together, suggesting that the earth, matter, is indeed composed of body together with light.

---

creations: the heavens (שָׁמַיִם) and the earth (אֶרֶץ), representing the spiritual and physical realms, the soul and the body.▲

Their initial letters spell "fire" (אֵשׁ), with the *alef* of "earth" (אֶרֶץ) preceding the *shin* of "heavens" (שָׁמַיִם).

Though in the Torah's first verse the heavens appear before the earth, the second account of creation begins with a parallel verse which concludes with the earth appearing before the heavens▶▶, "These are the generations of the heavens and the earth upon being created, on the day that *Havayah Elokim* made earth and heavens."[21]

The apparent contradiction regarding the order of creation of the heavens and the earth found its way into a disagreement between the academies of Hillel and Shamai. The academy of Shamai, based on the Torah's first verse, states that the creation of the heavens preceded that of the earth. Hillel's academy, based on the end of the second account's first verse, states that the creation of the earth preceded that of the heavens.

To reconcile the two seemingly conflicting orders, we must differentiate between thought and deed. The Torah's first verse describes the creation of the heavens and earth in deed. But, in

▶▶ Heavens first and then earth reflects the process of "**evolution**" (הִשְׁתַּלְשְׁלוּת) in creation, from the spiritual to the physical, but first earth and then heavens reflects creation *ex nihilo*, the "something," the earth, suddenly appearing from nothing that preceded it, without the more spiritual "heavens" to precede it.

---

21. Genesis 2:4.

thought, the creation of the earth preceded that of the heavens, for "the end of deed is first in thought."[22] The ultimate purpose of creation is to reveal God's absolute unity on earth and with this thought in mind God created the world. This is what is alluded to in the very first word of the Torah, the covenant of fire, with earth preceding heavens in the word fire. One may say that the order of earth before heavens (body before soul), the order adopted by Hillel's academy, is what fired the Creator in the act of creation. ◄

The third manifestation of the covenant of fire is thus the covenant of earth-heavens:

בְּרִית אֶרֶץ שָׁמַיִם

The fourth covenant of fire: One and Second

The first day of creation concludes, "… and it was evening and it was morning *one* day"▾▾ (וַיְהִי עֶרֶב וַיְהִי בֹקֶר יוֹם אֶחָד).[23] The second day concludes, "… and it was evening and it was morning, second day" (וַיְהִי עֶרֶב וַיְהִי בֹקֶר יוֹם שֵׁנִי). The first day ends with the word "one" (אֶחָד) and the second day with the word "second" (שֵׁנִי). The initial letters of these two words spell "fire" (אש).

In Kabbalah it is explained that "light" (אוֹר), "water" (מַיִם), and "firmament" (רָקִיעַ), the three things created on the first two days,

---

22. *Siddur, Kabalat Shabbat, Lecha Dodi.*
23. Genesis 1:5.

▶ Similar to the idea that thoughts about the earth were what motivated the Creator to create reality, we find the notion that (Psalms 85:12), "Truth shall spring out of the earth" (אֱמֶת מֵאֶרֶץ תִּצְמָח), i.e., that **a higher level of truth** would be produced by the earth than the one found in the heavens.

Above we saw that the word "earth" (אֶרֶץ) is an acronym for the phrase (Psalms 97:11), "light is sown for the righteous" (אוֹר זָרֻעַ לַצַּדִּיק). The second verse of the Torah describes the initial, chaotic state of the earth, and then comes the third verse, "And God said, 'Let there be light!' and there was light." The infinite light concealed in the earth (to be), the body (the feminine principle), is what fired God in the act of creation, *ex nihilo*.

▶▶ The accounts of all 6 days of creation end with the phrase, "**and it was evening and it was morning…**" (וַיְהִי עֶרֶב וַיְהִי בֹקֶר...). The *gematria* of these four words is itself a multiple of 6 (636 = 6 · 106), suggesting a connection with all six days.

But, more amazing is that in primordial numbering, their value is 2808 = 13 · 216. 13 is the value of "love" (אַהֲבָה) always associated with Abraham, who instituted the morning prayer, *Shacharit* and 216 is the value of "fear" (יִרְאָה) associated with Isaac who instituted the afternoon prayer, *Minchah*. Also, 216 = 6³.

Just the value of the two words, "and it was" (וַיְהִי) in primordial numbering is 292, a multiple of 73 (292 = 4 · 73), whereas the

value of the two words "evening… morning" (עֶרֶב בֹקֶר) is 2516, a multiple of 37 (2516 = 68 · 37), the two companion numbers whose product equals the first verse of the Torah.

The average value of the four words (וַיְהִי וַיְהִי עֶרֶב בֹקֶר) in primordial numbering is 2808/4 = 702 = 27 · 26, the value of "Shabbat" (שַׁבָּת), suggesting that there are four experiences of the sanctity of Shabbat in every weekday.

In the entire Bible, this phrase appears only at the conclusion of the 6 days of creation, and 6 times 2808 = 16848 = 13 · 1296, where 13 is the *gematria* of "one" (אֶחָד)—the purpose of time is to reveal that all is one—and 1296 = 36² = 6⁴.

encapsulate the heart of the process of creation.[24] The initials of these three things, light, water, and firmament (אוֹר מַיִם רָקִיעַ) spell the root of the verb meaning "to say" (א.מ.ר). This verb is the one used by the Torah to refer to the secret of God creating all of reality with "Ten Sayings" or "Ten Utterances" (עֲשָׂרָה מַאֲמָרוֹת),[25] for example, "And [God] **said**, ['Let there be light...']" (וַיֹּאמֶר [אֱלֹהִים יְהִי אוֹר...]). Light is the spiritual seed of creation, water the relatively physical seed, and firmament the fetus, from which each of the individual entities of the six days of creation was brought into being.

"One day" (יוֹם אֶחָד) is the day in which the singular consciousness in reality was "*Havayah* is one" (יְהוה אֶחָד). On "the second day" (יוֹם שֵׁנִי), other consciousnesses (of angels) were created.[26] One is unity while two is duality, the beginning of plurality. The union of "one," "the countable one" (אֶחָד הַמָּנוּי),[27] and "second" thus reflects a level of absolute unity, "the true one" (אֶחָד הָאֱמֶת)[28] that is above, and thereby able to unite apparent duality-plurality and that state of unity which seems to oppose plurality.

This is the fourth covenant of fire alluded to in the first word of the Torah, the covenant of "one"-"second,"▾ of one and two:

בְּרִית אֶחָד שֵׁנִי

---

24. *Etz Chaim* 39:5.
25. *Avot* 5:1.
26. *Bereisheet Rabbah* 1:3.
27. See *Pardes Rimonim* 4:5.
28. See *Tanya* ch. 35.

▸ Combining the first letters of "**one**" (אֶחָד) **and** "**second**" (שֵׁנִי) spells "fire" (אֵשׁ). Combining the second letters spells "grace" or "favor" (חֵן), alluding to the final verse of the Torah portion of *Bereisheet* (Genesis 6:8), "And Noah found favor in the eyes of *Havayah*" (וְנֹחַ מָצָא חֵן בְּעֵינֵי יְהוה). Combining the third letters spells "enough" (דַי), the secret of the Name Almighty (שַׁדַּי) that the sages read as short for "He said to His world 'enough'" (שֶׁאָמַר לְעוֹלָמוֹ דַּי), as discussed above.

Let us picture this beautiful phenomenon (first horizontally and then vertically):

ד	ח	א
י	נ	שׁ

שׁ	א
נ	ח
י	ד

The primordial value of the two 3 letter words (or three 2 letter words) is $1392 = 24 \cdot 58$, where 58 is the value of "grace" (חֵן), the word formed by the 2 middle letters.

This word, "grace," appears 66 times in the Bible (without prefix letters): 26 times in the Pentateuch, 16 times in the Prophets, and 24 times in the Writings. The sum of the squares of these three numbers, $26^2 + 16^2 + 24^2 = 1508 = 26 \cdot 58$; 26 is the value of God's essential Name, *Havayah* and 58 is once again the value of

▷▷▷

The fifth covenant of fire: Adam and Seth

The initial "one" and "two" of creation ("one day" and "second day") also allude to the first and second generations of mankind, Adam (אָדָם) and Seth (שֵׁת), whose initials spell "fire" (אֵשׁ). Numerically, "one" (אֶחָד), 13, and "second" (שֵׁנִי), 360, equal 373, the midpoint of 745, which is the sum of "Adam" (אָדָם), 45, and "Seth" (שֵׁת), 700! The first two words of the Book of Chronicles are: "Adam Seth" (אָדָם שֵׁת).◄

There are two words that figure prominently in the description of both Adam's creation and Seth's birth. They are image (צֶלֶם) and likeness (דְּמוּת), referring to the inner and outer dimensions of reality respectively. The Torah describes how the Almighty created Adam in "the image of God" (בְּצֶלֶם אֱ-לֹהִים):[29]

וַיֹּאמֶר אֱ-לֹהִים נַעֲשֶׂה אָדָם בְּצַלְמֵנוּ כִּדְמוּתֵנוּ... וַיִּבְרָא אֱ-לֹהִים אֶת הָאָדָם בְּצַלְמוֹ בְּצֶלֶם אֱ-לֹהִים בָּרָא אֹתוֹ זָכָר וּנְקֵבָה בָּרָא אֹתָם

God said, "Let us make man in Our image as Our likeness"... God created man in His image, in the image of God He created him, male and female He created them.

29. Genesis 1:26-27.

▶ The first, middle, and final letters of "**Adam Seth**" (אָדָם שֵׁת) spell the word "truth" (אֱמֶת), alluding to the verse (Psalms 85:12), "Truth shall spring out of the earth." When the value of the final mem (ם) is set equal to 600 (as it is in major-numbering), then the words "Adam Seth" (אָדָם שֵׁת) equal 1305, also the value of the phrase (see *Vayikra Rabbah* 9:3), "The way of the earth precedes the Torah" (דֶּרֶךְ אֶרֶץ קָדְמָה לַתּוֹרָה). 1305 is also the product of 45 and 29, where 45 is the value of Adam (אָדָם), while 29 is half the value of Noah (נֹחַ). The idea here is that Adam –whose name sounds like the first two words, "the [way of the] earth precedes" (אֶרֶץ קָדְמָה)—precedes Seth, who represents Torah.

In addition, 1305 is the value of "woman" (אִשָּׁה), when the alef (א) is set equal to 1000, thereby alluding to the fact that Seth had to be born of woman, as in the verse, "A woman conceives and gives birth to a male [child]" (Leviticus 12:2). In fact, the value of these words in Hebrew is exactly 1305!

▷▷▷ "grace." Multiplying 26 by 58 is a direct allusion to the 26 instances of the word "grace" in the Pentateuch.

The *gematria* of the filling of the three words, "fire," "grace," and "enough" (אלף שין חית נון דלת יוד) is 1449 = 7 · 207, with 207 being the value of "light" (אוֹר). The second filling (אלף למד פא שין יוד נון חית יוד תו נון וו נון דלת למד תו יוד וו דלת) is 3200, the double square of 40, the *gematria* of the letter mem, the first and last letter in "water" (מַיִם).

The second filling of the first two words alone, "fire" and "grace" is 1820, considered by some to be the most important number in the Torah, as it is the number of times that God's essential Name, *Havayah* (יהוה) appears in the Torah.

We have now found the three levels of value for these three words: their normative value (373), their first filling (1449), and their second

filling (3200). Constructing a quadratic series from these values, we find that the next number in this series is, 5626,

373	1449	3200	5626
1076	1751	2426	
675	675		

The sum of the first four numbers of the series is 10648 = 22³, 22 being of course the letters of the *alefbeit*!

A final note about the the Hebrew word חֵן, normally translated as "grace" or "favor" that we started with. It is one of Hebrew's eight synonyms for "beauty," which correspond to the *sefirot* from understanding to kingdom. חֵן in particular denotes symmetric beauty and is the type corresponding with the *sefirah* of kingdom (מַלְכוּת), the feminine principle in Kabbalah. (For more on these 8 synonyms, see *The Art of Education*, pp. 246ff.)

At the same time, Seth is described as having been born in Adam's likeness,[30]

וַיְחִי אָדָם שְׁלשִׁים וּמְאַת שָׁנָה וַיּוֹלֶד בִּדְמוּתוֹ כְּצַלְמוֹ וַיִּקְרָא אֶת שְׁמוֹ שֵׁת

Adam lived one-hundred and thirty years; he begat a son in his likeness as his image; he called his name Seth.

It is only in describing the creation of Adam and the birth of Seth that the Torah idiomatically juxtaposes these two terms "image" and "likeness"▾: "in Our image as Our likeness" (בְּצַלְמֵנוּ כִּדְמוּתֵנוּ) for Adam, and "in his likeness as his image" (בִּדְמוּתוֹ כְּצַלְמוֹ) for Seth. But, there is of course an immediately apparent difference. With Adam the image (צֶלֶם) is predominant and the likeness (דְּמוּת) is subordinate—first "in Our image" (בְּצַלְמֵנוּ) then "as Our likeness" (כִּדְמוּתֵנוּ)—with Seth, the order is reversed with the likeness being predominant and the image being subordinate—first "in his [Adam's] likeness" (בִּדְמוּתוֹ) then "as his image" (כְּצַלְמוֹ).

---

30. Ibid. 5:3.

▶ In Kabbalah (*Sefer Halikutim, parashat Ha'azinu*), we are taught that "image," a masculine noun, and "likeness," a feminine noun, are relatively masculine and feminine. **Every individual possesses both a masculine and a feminine side**, as is apparent in the word "man" (אָדָם)—the first letter, the *alef* (א), represents the masculine, "image" aspect and the two final letter, *dalet* and *mem* (דמ), are the first two letters of "likeness" (דְּמוּת), represent the feminine aspect. The value of "image" (צֶלֶם) is equal to that of "tree" (עֵץ), alluding to the Tree of Life, meaning that "likeness" alludes to the Tree of Knowledge. The sum of all 5 elements, "man, image, likeness, Tree of Life, and Tree of Knowledge" (אָדָם צֶלֶם דְּמוּת עֵץ הַחַיִּים עֵץ הַדַּעַת טוֹב וָרָע) is 1820, both the number of instances that God's essential Name, *Havayah*, appears in the Pentateuch and also 10 times the value of "Jacob" (יַעֲקֹב), about whom the sages say that his face was strikingly similar to Adam's.

Still, relative to one another, Adam is primarily masculine while Seth is primarily feminine. This strengthens the significance of the covenant between them because as explained in the opening sections of this chapter, all covenants are in essence covenants of marriage between a masculine element and a feminine element.

In Genesis, chapter 5, the Torah enumerates the lifespan and milestones of the generations of mankind from Adam to Noah. The chapter begins with the verse, "This is the book of the chronicles of Adam, on the day God created man, He made him in the likeness of God." But, note that in this instance the Torah does not mention Adam's Divine image indicating that when describing mankind's timeline, only likeness, the relatively feminine half of the image-likeness pair of traits, is essential.

We learn from this that *likeness* (דְּמוּת) is linked with the passage of time, with an awareness of what physicists call the arrow of time (from past to future). This is our primary state of temporal consciousness after the primordial sin (instigated by Eve, hence the identification of likeness with feminine consciousness). On the other hand, *image* (צֶלֶם) is linked with time's eternal aspect, an awareness of eternal life (חַיִּים נִצְחִיִּים) the state of human consciousness before the primordial sin.

▶ The initial letters of "**a living soul**" (נִשְׁמַת חַיִּים) spell "Noah" (נֹחַ) whom the Torah describes (Genesis 6:8) as having found "favor" (חֵן, the reverse of נֹחַ) in the eyes of God.

Its final letters spell "sincere [man]" (תָּם), alluding to Jacob, whom the Torah describes (Ibid. 25:27) as a "sincere man" (אִישׁ תָּם). Noah too is described similarly as a "complete [from the same root as 'sincere'], righteous man" (אִישׁ צַדִּיק תָּמִים).

The middle letters (נִשְׁמַת חַיִּים) equal 360, also the value of (Job 31:2) "a portion of God from on high" (חֵלֶק אֱ-לוֹהַּ מִמַּעַל), a connotation for the Divine soul of Israel (see *Tanya*, beginning of ch. 2), the soul of Mashiach.

Considering the complete phrase in the Torah, "He blew into his nostrils a living soul" (וַיִּפַּח בְּאַפָּיו נִשְׁמַת חַיִּים), the value of the middle letters is 541, "Israel" (יִשְׂרָאֵל). The initial letters equal "son of David" (בֶּן דָּוִד), the Mashiach, and the final letters (וַיִּפַּח בְּאַפָּיו נִשְׁמַת חַיִּים) spell the word "seal" (חוֹתָם). This teaches us that the soul of the son of David, the Mashiach, is sealed in the Divine soul of Israel.

▶▶ The *gematria* of "**Yemen**" (תֵּימָן) is 500. The *gematria* of the more common word for "south" (דָּרוֹם) is 250, half of 500. Thus the two words exhibit the principle of "a whole and a half," the numerical meaning hidden in God's Name *Kah* (יָהּ).

In both cases, the first phrase begins, "in…" (...בְּ), the secret of the "transparent pane" (Moses' unique level of prophecy) and the second begins with, "as…" (...כְּ), the secret of the "translucent pane"—the level of the prophecy ascribed to all other prophets but Moses).

Let's look at the idiom in which the image appears for both Adam and Seth. Adam was created "in the image of God" (בְּצֶלֶם אֱ-לֹהִים) whose value is 248. Seth was created "in Adam's likeness" (בִּדְמוּת אָדָם) whose *gematria* is 497 (2 times 248 plus 1). Incredibly, the sum of the two idioms is 745, the value of "Adam Seth" (אָדָם שֵׁת)!

745 is also the value of the phrase, "the spirit in our nostrils, the Mashiach of God" (רוּחַ אַפֵּינוּ מְשִׁיחַ יְהוָה).[31] After forming the body of Adam from the dust of the earth, "[God] blew into his nostrils a living soul" (וַיִּפַּח בְּאַפָּיו נִשְׁמַת חַיִּים).[32] The two-letter root of "soul" (נִשְׁמַת) is "name," (שֵׁם) and the two-letter root of "living" (חַיִּים) is "alive" (חַי). Together, the two roots of "a living soul" spell "a living name" (שֵׁם חַי), a permutation of Mashiach (מָשִׁיחַ). Thus, the living soul that God blew into Adam's nostrils was the soul of Mashiach. The remaining letters of "a living soul" (נִשְׁמַת חַיִּים) permute to spell "Yemen" (תֵּימָן), synonymous in Hebrew with "south," ◀◀ as in the verse, "Awaken north wind and come south wind, blow upon my garden…."▼▼▼[33] The *gematria* of (נִשְׁמַת חַיִּים) is 858 = 33 times 26, *Havayah*.

---

31. Lamentations 4:20.
32. Genesis 2:7.
33. Song of Songs 4:16.

▶▶▶ In the verse, "**Awaken north wind and come south wind, blow upon my garden**," the garden symbolizes the body of man, the north wind awakens the desire in God to enliven man with the soul of Mashiach, and the south wind blows the soul of Mashiach into the nostrils of man.

Thus, "a living soul" (נִשְׁמַת חַיִּים) permutes to spell "Mashiach of the south" (מָשִׁיחַ תֵּימָן), whose soul is blown into his nostrils from the south wind, the direction and wind that corresponds to wisdom, as the sages teach that he who desires wisdom shall turn to the south.

▸▸ The first verse in Chronicles is, "**Adam Seth Enosh**" (אָדָם שֵׁת אֱנוֹשׁ). Note that the name of the third generation, Enosh, begins with an *alef* and ends with a *shin* (אֱנוֹשׁ), spelling "fire" (אֵשׁ). In fact, the entire verse begins with an *alef* and ends with a *shin*. The two fires in this verse allude to the fire in "man" (אִישׁ) and the fire in "woman" (אִשָּׁה). The value of the two sets of *alef shin* together is 602 = 7 · 86, or *Elokim* (אֱלֹהִים). The value of the entire verse is 1102 = 19 · 58, where 19 is the value of "Eve" (חַוָּה) and 58 the value of "Noah" (נֹחַ), so the remaining letters (apart from the two sets of *alef shin*) is 500, the gematria

of (Genesis 1:28) "Be fruitful and multiply" (פְּרוּ וּרְבוּ).

When the final *mem* of Adam is set equal to 600, the value of the entire verse becomes 1662 = 2 · 831. 831 is the sum of "*Elokim*" (אֱלֹהִים), 86, "Adam" (אָדָם), 45, and "Seth" (שֵׁת), 700! 831 is also the product of 3 and 277, meaning that the average value of each word is "seed" (זֶרַע) as in the words Eve used to described Seth's birth, "for God has set me another seed in place of Abel whom Cain slew." This also means that 1662 (twice 831) is equal to 6 · 277, or the value of all 6 permutations of "seed" (זרע זער רעז רזע עזר ערז).

▸ The idiom, "**The spirit of our nostrils**" (רוּחַ אַפֵּינוּ) equals 361 = 19^2; 19 is the value of "Eve" (חַוָּה), Adam's wife. Adding the value of "Adam" (אָדָם), 45, their sum is 406 = 7 · 58 = △28.

The sum of "spirit" (רוּחַ) and "Adam" (אָדָם) is 259 = 7 · 37, where 37 is the value of "Abel" (הֶבֶל). Seth (שֵׁת), 700, and Mashiach (מָשִׁיחַ), 358, equal 1058, the double square of 23, i.e., 2 · 23^2 (just 23^2 = 529, the gematria of "pleasure" [תַּעֲנוּג]).

▸▸▸ In Kabbalah, it is taught that the three letters of the name **Adam** (אָדָם) stand for Adam David Mashiach (אָדָם דָּוִד מָשִׁיחַ). The root of Adam (אָדָם) means "red" (אָדֹם), and David is called (1 Samuel 16:12) "the red one" (אַדְמוֹנִי). The primordial value of Adam (א אבגדהוזחטיכלמם) is 156, the product of 6 and 26. 156 is also the gematria of "Joseph" (יוֹסֵף). And so, the Mashiach contained in Adam possesses spiritual aspects from both the soul of Mashiach, the son of Joseph and Mashiach, the son of David.

The name **Seth** means "foundation" (See Bamidbar Rabbah 12:4) the origin in particular of the soul of Mashiach, the son of Joseph, known as "the righteous one, the foundation of the world" (Proverbs 10:25). The gematria of Seth (שֵׁת) is 700 = 50 · 14, and 14 is the value of "David" (דָּוִד), illustrating that Seth also includes the origin of the soul of Mashiach, the son of David.

In each of the two phrases—"the spirit in our nostrils▸, the Mashiach of *Havayah*" and "He blew into his nostrils a living soul" — there is one *alef* (the initial letter of Adam) and one *shin* (the initial letter of Seth), thus, the two phrases contains two explicit letter combinations of "fire" (אֵשׁ).▸▲ The *gematria* of the remaining letters in the two phrases is 1204 = 4 · 301, the value of four additional "fires" (אֵשׁ), so all together they contain a clear allusion to the six fires of creation! God blows a living soul, the soul of Mashiach, into the nostrils of Adam, created in His image, similar to His likeness, and Seth, begot from Adam in his likeness as his image, inherits that soul and becomes one with it.▸▸▸

Had Adam not sinned there would have been no need for "the way of the land" as a spiritual objective in and of itself.[34] But after the sin, Adam's first task became refining his character and, idiomatically, walking "the way of the land." When Seth was born (130 years later) "in his [Adam's] likeness as his image," the covenant of the Torah (whereby a father devotedly teaches his son Torah) was first established. This is the Torah implied in the second half of the idiom, "The way of the land precedes the Torah."

---

34. See *Sefat Emet Likutim, Tana Devei Eliyahu*.

▶ The three letters of the word "stone" (אֶבֶן) may further be interpreted to stand for **father, son, grandson** (אָב בֵּן נֶכֶד), three generations. There is another word, which in Biblical Hebrew means "son," נִין, pronounced *neen* (Genesis 21:23). *Neen* specifically refers to a son who is heir to the throne. In modern Hebrew it is used in reference to a great-grandson, the fourth generation.

The stone is a symbol for materiality and another acronym for "stone" (אֶבֶן) that is particularly related to father, son, and grandson is "land, house, and possessions" (אֲדָמָה בַּיִת נְכָסִים). The father purchases the land, the son builds a house on the land, and the grandson endeavors to fill the house with wealth and possessions.

The *gematria* of father, son, and grandson, when spelled in full (אלף בית בית נון נון כף דלת) is $1681 = 41^2$, where 41 is the value of "mother" (אם), revealing that it is actually the mother that inspires her son to continue his father's work and to actualize his potential. The mother serves as the link between the generations. All three—the land, the house, and the possessions—are in her merit.

▶▶ In the phrase, "From there he shepherd's the stone of Israel," the *gematria* of the word "**from there**" (מִשָּׁם) is 380, the same value as "Egypt" (מִצְרַיִם), the place to which it alludes.

▶▶▶ The *gematria* of the phrase, "From there does he shepherd the stone of Israel" (מִשָּׁם רֹעֶה אֶבֶן יִשְׂרָאֵל) is 1249. If we move the decimal point in this number so that it reads 12.49, we find that this is incredibly almost the **perfect root of 156**, the value of "Joseph" (יוֹסֵף)!

▶▶▶▶ The sum of "a snake upon a rock" (נָחָשׁ עֲלֵי צוּר), 764, and "stone of Israel" (אֶבֶן יִשְׂרָאֵל), 594 is 1358, the *gematria* of the second verse of the *Shema*: "Blessed be the Name of the glory of His kingdom for ever and ever" (בָּרוּךְ שֵׁם כְּבוֹד מַלְכוּתוֹ לְעוֹלָם וָעֶד).

▶▶▶▶▶ The value of "**stone of foundation**" (אֶבֶן שְׁתִיָּה) is 768, a multiple of 256, the *gematria* of Aharon (אַהֲרֹן) who as High Priest entered the Holy of Holies in the Temple, the site of the stone of foundation, once a year on *Yom Kippur*.

The covenant of Adam and Seth is the covenant of father and son. There is one simple three letter word in Hebrew interpreted by the sages as being short for father-son: "stone" (אֶבֶן),[35] whose first and second letters spell "father" (אָב), and whose second and third letters spell "son" (בֵּן)▲, with the second letter, the *beit* of "stone" (אֶבֶן), the first letter of the Torah, the letter common to both, uniting the two.

Based on this identification, the sages read the phrase from Jacob's blessing to his son Joseph,[36] "From there◀◀ does he shepherd the stone of Israel"◀◀◀ as meaning that in his capacity as the viceroy of Egypt, Joseph, like a shepherd, sustained his father and his father's sons (Joseph's siblings).

In the previous chapter, we saw that "the way of a snake upon a rock" corresponds to the element of earth and to the people of Israel.◀◀◀◀ Here we see that the stone too is a symbol for the people of Israel.

Indeed, the world was created from a stone, the stone upon which the Ark of the Covenant was placed in the Holy Temple in Jerusalem. That stone is known as the "foundation stone"◀◀◀◀◀ (אֶבֶן

---

35. *Rashi* on Genesis 49:24.
36. Ibid. 49:24.

שְׁתִיָּה), because, "from it the world was founded."[►37] The word for "foundation" (שְׁתִיָּה) indeed comes from "Seth" (שֵׁת), the foundation of mankind similarly because, "from him the world was founded."[38]

Interestingly, the initial letters of the "stone of foundation" (אֶבֶן שְׁתִיָּה) also spell "fire" (אֵשׁ), so the covenant of the stone of foundation can also be considered a covenant of fire. However, the covenant of the stone of foundation is not made up of two words that display a covenant between them, so it cannot be counted independently as one of our covenants here, but is mentioned only in merit of the covenant between Adam and Seth.

Another such "fire," indeed a pair of complementary terms, but included as well in the pair Adam Seth (father and son), is the pair: "father/s" (אָבוֹת) and "tribe/s" (שְׁבָטִים)[►►], whose initials spell "fire" (אֵשׁ). In Kabbalah, it is explained that the souls of the fathers, the Patriarchs, are from the World of Emanation, the world of pure Divine consciousness, whereas the souls of tribes (the sons of Jacob, the third and "choice" of the Patriarchs), excluding Joseph, are from the World of Creation, the world where souls begin to become conscious of themselves as independent created beings.

To conclude, the fifth covenant of fire alluded to in the first word of the Torah is the covenant between father and son, the first father and the first son that was like his father (in contrast to Cain and Abel who were not like their father[39]), the covenant of the chain of generations from the beginning of creation, the covenant of Adam and Seth:

$$\text{בְּרִית אָדָם שֵׁת}$$

▶ The sages use the idiom, "**from it/him the world was founded**" (שֶׁמִּמֶּנּוּ הוּשְׁתַּת הָעוֹלָם) in reference to both Seth and to the stone of foundation. Its value is 1698. When added twice (once for Seth, once for the stone of foundation) to the value of "Seth" (שֵׁת), which is 700, the sum comes to $4096 = 64^2 = 16^3 = 16 \cdot 256$, where 256 is the value of "Aaron" (אַהֲרֹן)! Adding, "the stone of foundation" (אֶבֶן שְׁתִיָּה), 768, the total sum becomes $4864 = 19 \cdot 256$, where 19 is the value of Eve (חַוָּה) and the reduced value of "wisdom."

▶▶ The pair of "**father**" and "**tribes**" is beautifully seen on the Jewish calendar. The name of the fifth month (the months are counted from *Nisan*) is Av (אָב), which literally means "father." The eleventh month is *Shevat* (שְׁבָט), whose literal meaning is "tribe." Organizing the 12 months on a circle, we find these two months exactly opposite one another.

37. *Vayikra Rabbah* 20:4.
38. *Bamidbar Rabbah* 14:11.
39. See the commentaries on Genesis 5:3.

The sixth covenant of fire: Abraham and Sarah

The sixth and final pair that we will contemplate is the first Jewish couple, Abraham (אַבְרָהָם) and Sarah (שָׂרָה), whose initials spell "fire" (אֵשׁ). This is the only explicit marital union among the six covenants of fire we have meditated upon.

Of the four elements fire, air, water, and earth, Abraham, the first Jew corresponds in particular to the element of water, the element that symbolizes love and kindness, the attribute of Abraham, as in the words of the prophet,[40] "loving-kindness to Abraham" (חֶסֶד לְאַבְרָהָם). In our prayers (for rain, water) we ask God to "remember the father [Abraham] who was drawn after You like water." ◂

The nature of water is to descend from on high to the lowest possible point of rest.[41] Water flows down from its spiritual source (the Divine attribute of loving-kindness) to impregnate the physical realm with Divine consciousness.[42] This was the spiritual life-work of Abraham. Until his time, God was only "the God of the heavens" (אֱ-לֹהֵי הַשָּׁמַיִם),[43] ▾▾ that is to say that His presence was only recognized

▸ The average value of the five words, "remember the father [Abraham] who was drawn after You like water" זְכוֹר אָב נִמְשַׁךְ אַחֲרֶיךָ כַּמַּיִם) is 199, the gematria of "charity" (צְדָקָה), Abraham's primary form of Divine service (see Genesis 15:6).

---

40. Micah 7:20.
41. See Tanya ch. 4.
42. In Kabbalah, the male and female seeds are called "masculine waters" and "feminine waters."
43. Genesis 24:7. See Rashi.

▸▸ The gematria of "**God of the heavens**" (אֱ-לֹהֵי הַשָּׁמַיִם) is 441 = 21^2, also the value of "truth" (אֱמֶת).

The Torah, God's truth, descends from heaven (Exodus 20:18). The average value of the 9 letters that compose the idiom "God of the heavens" is 49 = 7^2, alluding to the inter-inclusion of the 7 emotive attributes with one another.

Since the phrase has 9 letters we can depict it as a 3 by 3 square and focus on the four corners and middle point

$$\begin{array}{ccc} א & ל & ה \\ י & ה & שׁ \\ מ & י & ם \end{array}$$

(This is the same as taking every other letter in the phrase). The value of the highlighted letters, אההמם, is 91 = △13. 91 is also the value of the union of God's essential Name, Havayah (יהוה) and the Name used to pronounce it, Adni (אֲדֹנָי). It is also the value of the word "Amen" (אָמֵן).

The value of the complete phrase in primordial numbering is 1631 = 7 · 233, where 233 is the gematria of "the Tree of Life" (עֵץ הַחַיִּים).

Adding these two values of the phrase together, 441 ⊥ 1631 = 2072 = 7 · 296, the gematria of "the earth" (הָאָרֶץ), implying that the higher earth, the Garden of Eden, with the Tree of Life in its middle, is present in the "God of the heavens." As such, it is always ready to descend, by means of Abraham's spiritual path—the path of loving-kindness—from heaven, to manifest itself in the lower realm, the lower earth!

▶ The initials of "God of the heavens and God of the earth" (אֱ־לֹהֵי הַשָּׁמַיִם וֵא־לֹהֵי הָאָרֶץ) spell **God's hidden Name, *Ekveh*** (אהוה), the same Name that connects the heavens to the earth in the Torah's first verse by appearing in their initials (אֵת הַשָּׁמַיִם וְאֵת הָאָרֶץ). Because its value is 17, the *gematria* of "good" (טוֹב), the Name *Ekveh* is referred to as "the goodly Name."

Like glue, good makes things cling together. In fact, glue is described as being good in the verse, "say to glue, it is good" (Isaiah 41:7).

Clinging describes the union of a married couple: "He shall cling to his wife and they shall become one flesh" (Genesis 2:24) and indeed the relationship between them is also described as "good": "One who has found a wife has found good" (Proverbs 18:22).

Adding the value of "the heavens and the earth" (אֵת הַשָּׁמַיִם וְאֵת הָאָרֶץ) to "God of the heavens and God of the earth" (אֱ־לֹהֵי הַשָּׁמַיִם וֵא־לֹהֵי הָאָרֶץ) gives 2288 = 88 · 26. The average value of the 8 words in both phrases is 286 = 11 · 26 = 13 · 22, or "one" (אֶחָד) times "together" (יַחַד).

---

by heavenly states of consciousness (i.e., angels), but Abraham made Him "the God of the earth" (אֱ־לֹהֵי הָאָרֶץ),[44] he brought Him into the consciousness of earthly beings.▲

Sarah, his wife, converts the innate waters of Abraham to fire (as alluded to by the combination of the initial letters of their names that spell "fire"). As we saw above, God creates fire with the letter *shin*, the initial letter of Sarah. In general, the fire of woman (אִשָּׁה) is stronger (i.e., more intrinsic to and predominant in her character) than that of man (אִישׁ).

God says to Abraham,[45] "all that Sarah says to you, listen to her voice" (כֹּל אֲשֶׁר תֹּאמַר אֵלֶיךָ שָׂרָה שְׁמַע בְּקֹלָהּ). From this verse the sages learn that "Abraham was subordinate to Sarah in prophecy" (שֶׁהָיָה אַבְרָהָם טָפֵל לְשָׂרָה בִּנְבִיאוּת).▼▼[46] Abraham is instructed to listen to and abide by

---

44. Ibid. 24:3.
45. Genesis 21:12.
46. *Midrash Tanchuma, Shemot* 1.

▶▶ The *gematria* of the word used here for "**prophecy**" (נְבִיאוּת) is 469 = 7 · 67, "understanding" (בִּינָה), the mother principle—Sarah is the first matriarch of the Jewish people.

469 is the number of words in the seven days of the first account of creation (it is also the 13th Shabbat/hexagonal number; see page 53).

469 is also the sum of "pleasure" (עֹנֶג), 123, and "will" (רָצוֹן), 346, the two essential dimensions of the crown, suggesting that prophecy depends on uniting these two parts of the super-conscious.

Pleasure is the origin in the soul of the element of water and will is

the origin of the element of fire. In their union water converts to fire, Abraham converts to Sarah as it were, and so we understand why he is subordinate to her in prophecy.

The complete saying of the sages, "All that Sarah says to you, listen to her voice' from here you learn that Abraham was subordinate to Sarah in prophecy" (כֹּל אֲשֶׁר תֹּאמַר אֵלֶיךָ שָׂרָה שְׁמַע בְּקֹלָהּ, מִכָּאן אַתָּה לָמֵד שֶׁהָיָה אַבְרָהָם טָפֵל לְשָׂרָה בִּנְבִיאוּת) = 4589 = 13 · 353, or "love" (אַהֲבָה) times "joy" (שִׂמְחָה), which is particularly related to our Matriarchs, as in the verse, "The mother of children is joyful" (Psalms 113:9).

／

▶ The sum of **Abraham** (אַבְרָהָם) and "**water**" (מַיִם) is 338, or 26 · 13, the double square of 13, alluding to the verses, "Abraham was one" (Ezekiel 33:24) and, "I have called him 'one'" (Isaiah 51:2).

The sum of Sarah (שָׂרָה) and "fire" (אֵשׁ) is 806, or 26 times 31 (the reverse of 13).

Together the two pairs equal 44 · 26, or 88 · 13, and since in all 4 words there are 13 letters, this means that the average value of each letter is 88. These 13 letters divide in a golden division (following the Love-Fibonacci series of numbers) into 8 (אַבְרָהָם מַיִם) and 5 (שָׂרָה אֵשׁ).

Within each pair, the letters further divide in a golden division again—the 8 into 5 and 3, the 5 into 3 and 2!

If we fill the letters of the four words (אלף בית ריש הא מם מם יוד מם שין ריש הא אלף שין), their value becomes 2646, or 6 times the value of "truth" (אֱמֶת), meaning that 2646 is equal to the value of all 6 permutations of "truth" (אמת אתם מתא מאת תאם תמא), or 6 · 41. 2646 is also equal to 7 · 378, the value of "chashmal" (חַשְׁמַל), the Bible's most mysterious word. 378 is also the triangle of 27 = 3³.

The primordial value of these 4 words is 4305 = 5 · 861, where 861 is the triangle of 41 and equal to 7 times "pleasure" (עֹנֶג).

▶▶ The *gematria* of the two roots "**say**" (אמר) **and** "**listen**" (שמע) is 651 = 21 · 31, the product of two of God's Names, *Ekyeh* (אֶהְיֶה), 21—the Name corresponding to the mother principle (the *sefirah* of understanding) and representing Sarah—and *Kel* (אֵל), 31—the Name corresponding to loving-kindness, the attribute of Abraham. Just as understanding is higher than loving-kindness (in the order of emanation) so in this case, Sarah is higher than Abraham, as if she is the mother and he is the son.

These two roots' primordial value is 2506 = 7 · 358, the *gematria* of Mashiach (מָשִׁיחַ). The primordial component alone, which is actually the difference between the primordial value (2506) and the normative value (651) is 1855 = 5 · 371, where 371 is the union of "nothing" (אַיִן) and "something" (יֵשׁ) seen earlier.

the words of Sarah, to take her advise, to follow her prophetic intuition. His water is told to become one with her fire.▲

The initials of the two 3-letter verbs in the above phrase, "all that Sarah *says* to you, *listen* to her voice"—"say" (אמר) and "listen" (שמע)◀◀—spell "fire" (אֵשׁ). The second letter of each root is *mem*, the letter of water, indicating that water, Abraham's innate element, is indeed subordinate to and included within fire, Sarah's innate element.

The phrase "And Abraham said" (וַיֹּאמֶר אַבְרָהָם) appears 7 times in the Torah (each time at the beginning of a verse, opening a new state of consciousness).[47] The *gematria* of the two words is 505, or "Sarah" (שָׂרָה)! The soul of Sarah, the source of prophetic inspiration, speaks through the mouth of her husband Abraham, just as the Divine Presence, the *Shechinah* (שְׁכִינָה), the feminine dimension of Godliness, speaks from the throat of Moses.[48] The three letters of the root "say" (אמר) stand for the three elements fire, water, air (אֵשׁ מַיִם רוּחַ), in that order. The predominant element, the first letter of the root, is the fire of Sarah, followed by the water of Abraham and finally the air (vapor) of the spoken word.

In Psalms we find,[49]

---

47. All 7 instances are in Genesis: 17, 18; 20, 2; 20, 11; 21, 24; 22, 5; 22, 8; 24, 2.
48. See *Zohar* III, 219a.
49. 39:4.

חַם לִבִּי בְּקִרְבִּי בַּהֲגִיגִי תִבְעַר אֵשׁ דִּבַּרְתִּי בִּלְשׁוֹנִי

My heart is hot within me, as I contemplate a fire shall burn, I spoke with my tongue.▸

Speech begins with fire, with the flame of the heart, with the soul-root of Sarah.

The initial letters of the three phrases of the verse (חַם לִבִּי בְּקִרְבִּי, בַּהֲגִיגִי▸▸ תִבְעַר אֵשׁ, דִּבַּרְתִּי בִּלְשׁוֹנִי) spell *chabad* (חַבַּד), the acronym of the three intellectual *sefirot*—wisdom, understanding, and knowledge.

In this context, these are the three intellectual faculties of the heart itself, in contrast with those of the brain. These are the three intellectual faculties of Sarah, the heart, the element of fire, in contrast with those of Abraham, symbolizing the intelligence of the brain.

The highest level of the soul, the "single one" (יְחִידָה), with its own unique intelligence, resides in the heart, not in the brain. The first word of the first phrase, "hot" (חַם) spells "brain" (מֹחַ) backwards. In the *Zohar's* introduction it states that, "wisdom is the brain." The contemplation of the heart in the second phrase is the explicit property of understanding. The speaking of the tongue, the third phrase, is a property of knowledge.[50]

Another peculiarity arising from the intellectual faculties in this verse being those of the heart (Sarah) rather than the brain (Abraham) can be seen in the time tense used in each of the three phrases that make up the verse: "My heart is hot within me" is in the present, "as I contemplate a fire shall burn," is in the future, and "I spoke with my tongue" is in the past.

In Kabbalah, this order would usually imply the order of the three intellectual *sefirot*, knowledge (present), understanding (future), and wisdom (past), the reverse of the natural order of the *sefirot*, which is wisdom, understanding, and knowledge. This means that in the context of this verse, wisdom and knowledge exchange tenses—a

▸ This verse has 8 words and 33 letters. The value of verse's first letter, the *chet* of "hot" (חַם) is 8 provides self-reference to the number of words in the verse.

The verse's fourth word, "as I contemplate" (בַּהֲגִיגִי) illustrates self-reference: its value is 33, the number of letters in the complete verse. The total *gematria* of every other letter in this word (בגג) is 8, the number of words in the complete verse.

▸▸ The value of the fourth word, "**as I contemplate**" (בַּהֲגִיגִי), 33 is also the value of "heart" in Aramaic (לִבָּא) mentioned in the *Tikunei Zohar* in the well-known statement, "understanding is in the heart" (בִּינָה לִבָּא), referring to understanding that is part of the intellectual faculties in the heart.

---

50. See Isaiah 50:4.

151

phenomenon known as "exchanging places" (אַחְלִיפוּ דוּכְתַּיְיהוּ)— wisdom becomes present and knowledge past. From this we learn that wisdom of the heart experiences the present, while its knowledge knows the past.▼

The *gematria* of the first phrase of the verse, "My heart is hot within me" (חַם לִבִּי בְּקִרְבִּי), 404 = "holiness" (קֹדֶשׁ), interpreted by the sages as short for "on fire" (יְקוֹד אֵשׁ).[51] The *gematria* of the remaining two phrases of the verse, "as I contemplate a fire shall burn, I spoke with my tongue" (בַּהֲגִיגִי תִבְעַר אֵשׁ דִּבַּרְתִּי בִּלְשׁוֹנִי), 2020 = 5 · 404, the value of the first phrase! Thus the whole verse = 2424 = 6 · 404, 6 times "holiness," corresponding to the 6 permutations of the three letter word "holiness" (קֹדֶשׁ), and to the 6 fires of creation!

There are two verses in the Prophets in which Abraham is referred to as "one." Isaiah[52] says "for he was one when I called him◄◄" and

---

51. *Chulin* 115a.
52. 51:2.

▶▶ In Kabbalah, **"to call upon a soul"** implies drawing it down from its spiritual source, as in the verse (Isaiah 41:4), "He calls the generations from the beginning [literally, 'from the head']" (קֹרֵא הַדֹּרוֹת מֵרֹאשׁ), interpreted to mean that God draws down each generation of souls from its source in the head of Primordial Man (אָדָם קַדְמוֹן).

▶ The verse divides into three phrases: 1) חַם לִבִּי בְּקִרְבִּי, 2) בַּהֲגִיגִי תִבְעַר אֵשׁ, and 3) דִּבַּרְתִּי בִּלְשׁוֹנִי that possess 10, 12, and 11 letters respectively. The sum of the squares of these three consecutive integers is $10^2 + 11^2 + 12^2 = 365$, the number of prohibitive commandments in the Torah (expressing the attribute of might, fire), and days of the solar year, with the sun another symbol of might and fire).

365 can also be expressed as the sum of just two consecutive squares: $13^2 + 14^2$. So we have **the beautiful arithmetic identity**:

$$10^2 + 11^2 + 12^2 = 13^2 + 14^2$$

This dual representation with all 5 roots—10, 11, 12, 13, and 14—being consecutive is unique to 365 alone!

However, it is important to note that this unique property of 365 is part of a larger, amazing property of integers and squares.

Look at the following. It is true to infinity:

$$1 + 2 = 3$$
$$4 + 5 + 6 = 7 + 8$$
$$9 + 10 + 11 + 12 = 13 + 14 + 15$$

$$16 + 17 + 18 + 19 + 20 = 21 + 22 + 23 + 24$$

and so on.

Note that in this well-known relationship that the first numbers on the left-hand side are the square numbers, while the first numbers on the right-hand side are the covenant numbers, where the covenant number of $n = n^2 + n + 1$.

A similar property exists for square numbers (one line of which relates to 365, as we saw):

$$3^2 + 4^2 = 5^2$$
$$10^2 + 11^2 + 12^2 = 13^2 + 14^2$$
$$21^2 + 22^2 + 23^2 + 24^2 = 25^2 + 26^2 + 27^2$$
$$36^2 + 37^2 + 38^2 + 39^2 + 40^2 = 41^2 + 42^2 + 43^2 + 44^2$$

and so on.

The key to this amazing property of the squares is that the first numbers on the left-hand side are every other triangular number (3, 10, 21, 36, 55, …). The first numbers on the right-hand side are the interface, or inspirational numbers (5, 13, 25, 41, 61, …)! No similar property exists for cubes or other powers, making this one of the most beautiful properties that the Creator instilled in pure number theory!

Ezekiel[53] says "Abraham was one." The *Zohar*[54] states that the "one"▶ in the first verse refers to the union of Abraham and Sarah.

The complete verse from Isaiah reads,

Look unto Abraham your father and to Sarah that bore you, for he was one when I called him, and I blessed him, and I increased him.▶▶ ▼▼▼

In their source, Abraham and Sarah are one. God called them, blessed them, and increased them. These three verbs correspond to three stages of self-fulfillment (corresponding to the three spiritual worlds of Creation, Formation, and Action), called in mind, blessed in heart, and increased in action.

In another place, the *Zohar*[55] explains that to be *one*, one must be married, as the Torah says about the union of husband and wife, "and he shall cling to his wife and they shall become *one* flesh."[56]

And so the sixth and final covenant of fire of creation is the covenant of the first Jewish marriage, the covenant of Abraham Sarah, the covenant that converts water into fire:

בְּרִית אַבְרָהָם שָׂרָה

---

53. 33:24.
54. I, 22b.
55. III, 81a.
56. Genesis 2:24.

▶ In the word "one" (אֶחָד) itself the first two letters, *alef-chet*, "brother" (אָח) are its male dimension, and the final *dalet* is the female dimension.

▶▶ In Hebrew, the complete verse reads, הַבִּיטוּ אֶל אַבְרָהָם אֲבִיכֶם וְאֶל שָׂרָה תְּחוֹלֶלְכֶם כִּי אֶחָד קְרָאתִיו וַאֲבָרְכֵהוּ וְאַרְבֵּהוּ.

Analyzing the verse from a numerical perspective, we find that its key word is "your father" (אֲבִיכֶם) whose value, 73 is self-referential as it is also the value of "wisdom" (חָכְמָה), also called "the father principle" (אַבָּא). Of course the verse is about the patriarch Abraham.

We then discover that other parts of the verse are also multiples of 73. The *gematria* of "Look unto Abraham your father and to Sarah that bore you" (הַבִּיטוּ אֶל אַבְרָהָם אֲבִיכֶם וְאֶל שָׂרָה תְּחוֹלֶלְכֶם) is 1460 = 20 · 73.

"He was one when I called him" (אֶחָד קְרָאתִיו) equals 730 = 10 · 73, one half of 1460.

At the same time, in a beautiful example of inter-inclusion between Abraham and Sarah, the *gematria* of the entire verse is 2680 = 40 · 67, where 67 is the value of "understanding" (בִּינָה), also called "the mother principle" (אִמָּא), referring of course to Sarah.

▶▶▶ The root of the final word and culmination of the verse, "**and I increased him**" (וְאַרְבֵּהוּ) is *reish-beit-hei* (רבה), whose value is 207, the value of "light" (אוֹר), "infinity" (אֵין סוֹף).

Each of the root's three letters appears 5 times in the verse, meaning that their sum value is 5 · 207 = 1035, the triangle of 45, the value of "man" or "Adam" (אָדָם). Since 1035 is also equal to 5 times "light" (אוֹר), it also alludes to the 5 instances of "light" appearing in the Torah's account of the first day of creation.

The number 5 is the value of the letter *hei* (ה), the letter added to Abraham upon circumcision, changing his name from Avram (אַבְרָם) to Abraham (אַבְרָהָם).

These letters *reish-beit-hei* are the three middle letters of "Abraham" (אַבְרָהָם). "Sarah" shares the two letters *reish* and *hei* (שָׂרָה). The remaining three letters in their names are, *alef*, *mem*, and *shin*, the three "mother letters" of the Hebrew *alefbeit*, as explained in the Book of Formation. Their value is 341, the *gematria* of "the shield of Abraham" (מָגֵן אַבְרָהָם).

153

The six covenants of fire

Whenever our meditation on a particular concept yields a group of elements—in this case, 6 different covenants of fire—our meditation must be completed by understanding the relationship between the different elements in the group. Discovering the relationship between elements in a group is done by corresponding them to a particular Torah model. Performing this part of the meditation is the most difficult and requires in-depth experience with the manner in which the scholars of the past have done the same for other groups of elements. Once complete, the correspondence resulting from this effort yields what is called a *partzuf*, a model, in Kabbalah.

Of the six covenants of fire described in this chapter, only the covenant of *Kel Shakai* (בְּרִית אֵ־ל שַׁ־דָּי) is a covenant of two of God's Names. ◄

Although the juxtaposition of these two Names in the Torah is usually associated with the *sefirah* of foundation (as is the concept of covenant, in general), here the two Names, as a pair linked together in a covenant of marriage correspond to the two dimensions of the *sefirah* of crown. It is in the crown, the super-conscious level of the Divine soul of Israel, that God's Names manifest and unite (thereby rectifying the blemishes of the *sefirah* of foundation).

Kel Shakai is usually translated "God Almighty." In the terminology of Kabbalah, the Name *Kel* (standing for "God") denotes the source of loving-kindness in the inner persona (פַּרְצוּף) of the crown, the super-conscious pleasure principle of the soul, as it enters and becomes enclothed in the crown of the outer persona, super-conscious will. The Name *Shakai* (standing for "Almighty") denotes the might of the inner persona of the crown that enters and becomes enclothed in the wisdom of the outer persona. These are the two essential links, or "rectifications" (תִּקּוּנִים) that connect the two persona of the crown together. ◄◄

In Kabbalah these two rectifications are known as the "skull"

▶ Each of **God's Names** is a manifestation of His infinite light as it was before the initial contraction that begins the creative process.

▶▶ The inner persona of **the crown** is known as "Ancient One" (עַתִּיק) and its value is 580, and the outer persona is called the "Long One" (אֲרִיךְ) and its value is 231. The value of the two Names *Kel Shakai* (אֵ־ל שַׁדַּי) is 345, also the *gematria* of "Moses" (מֹשֶׁה). The sum of all 4 words (עַתִּיק אֲרִיךְ אֵ־ל שַׁדַּי) together is $1156 = 34^2$, which means that their average value is $289 = 17^2$, where 17 is the value of "good" (טוֹב), linking them once again to Moses, who at birth was called "good" (Exodus 2:2). 289 is also the value of "God created" (בָּרָא אֱ־לֹהִים), the second and third words of the Torah.

(גֻלְגַּלְתָּא) and the "concealed brain" (מוֹחָא סְתִימָאָה) respectively. The "skull" shines its brilliant light of love and kindness in the World of Emanation whereas the "concealed brain" shines its light to the three lower worlds of Creation, Formation, and Action.

The next covenant is the covenant of nothing and something (בְּרִית אַיִן יֵשׁ). This pair corresponds to the two *sefirot* wisdom and understanding▸. In Kabbalah, wisdom is referred to as "nothing" (אַיִן) and understanding as "something" (יֵשׁ).

As noted, wisdom and understanding are the father and mother principles, whose union is continuous for the sake of the continuous re-creation of all reality, created initially in seven days symbolizing the seven offspring of father and mother: 6 sons—the six work days— and 1 daughter—Shabbat, on which day God rested, i.e., created rest. In *Sefer Yetzirah* father and mother are called "the depth of beginning" (עֹמֶק רֵאשִׁית) and "the depth of end" (עֹמֶק אַחֲרִית) respectively. Similarly, the *alef* of nothing (אַיִן) is the first, the beginning letter while the *shin* of something (יֵשׁ) is the final, the end letter.

The next covenant is the covenant of earth and heavens▸▸, physicality and spirituality (בְּרִית אֶרֶץ שָׁמַיִם). This is the union of daughter and son or, in the context of marriage, bride and groom that follows the union of father and mother. These two unions correspond to the four letters of God's essential Name *Havayah*. The higher union of father and mother corresponds to the union of the first two letters of *Havayah*, *yud-hei* (י־ה). The lower union of bride and groom corresponds to the union of the two final letters of *Havayah*, *vav-hei* (ו־ה). In terms of the *sefirot*, the earth, the bride, corresponds to the *sefirah* of kingdom, *malchut* (מַלְכוּת), and the heavens, the groom, corresponds to the *sefirah* of beauty, *tiferet* (תִּפְאֶרֶת).

The fourth covenant is the covenant of one and two, "one day" (יוֹם אֶחָד) and "second day" (יוֹם שֵׁנִי) of creation (בְּרִית אֶחָד שֵׁנִי). The first day of creation corresponds to the *sefirah* of loving-kindness (חֶסֶד) and the second day of creation corresponds to the *sefirah* of might (גְּבוּרָה). The inner experiences of these two *sefirot* are love and fear,

▸ The sum of the values of "nothing" (אַיִן), 61, "something" (יֵשׁ), 310, "wisdom" (חָכְמָה), 73, and "understanding" (בִּינָה), 67, is 511, or 7 · 73, where 73 is again the value of "wisdom!"

Both Judaism and non-Jewish culture agree that there are **7 categories of wisdom** (although there is disparity in their identification). What this *gematria* reveals is that every true wisdom includes a union between nothing and something, permeating the wisdom and understanding within it.

▸▸ The heavens and the earth correspond to the two final letters of *Havayah* (וה); nothing and something correspond to the first two letters of *Havayah* (יה). Combining **the two covenants** together would give us the letters of *Havayah* in the order: yud-hei-hei-vav (יההו), with the final two letters in reverse order because in the covenant of earth and heavens, the earth precedes the heavens.

This order of letters in *Havayah* is considered the second of the twelve possible permutations of the Name. This permutation is the acronym for the order of the four Torah portions contained in *tefilin*, according to the opinion of Rabbeinu Tam.

the two spiritual hands of the soul, the right hand and the left hand, respectively. With His right hand, love, God creates the heavens and with His left hand, fear, God creates the earth as stated in Isaiah,[57] again with the earth preceding the heavens, as in the third covenant of creation.[58] So the fourth covenant corresponds to the *sefirot* loving-kindness and might.

The fifth covenant is the covenant of the first two generations of mankind, the covenant of Adam Seth (בְּרִית אָדָם שֵׁת). Adam, the first human being, represents consciousness in general, the *sefirah* of knowledge. He ate from the forbidden fruit of the Tree of Knowledge of Good and Evil. Had he waited until Shabbat the fruit would have become permissible[59] and the knowledge obtained thereby would have been pure consciousness of the Divine◄.

Had Adam withstood the test of not partaking from the forbidden fruit he would have become the first Jew. The essential attribute of the Divine soul of Israel is its consciousness, i.e., knowledge of God. Additionally, the union of Adam and Eve is referred to as knowing one another.[60]

The name Seth means foundation. From him the world was founded, as explained above. Clearly then he corresponds to the *sefirah* of foundation. Foundation in the heart is like knowledge in the mind. The function of both is to connect opposites, to connect masculine and feminine. Knowledge connects father (wisdom) with mother (understanding) and foundation connects the bride (kingdom) with the groom (beauty). And so the pair of the fifth covenant of fire corresponds to the pair of *sefirot* knowledge and foundation◄◄.

The next and final of the six covenants of fire in creation is the covenant of the first Jewish couple, Abraham (through whose soul-

▶ The **Ba'al Shem Tov** revealed that had Adam waited until Shabbat to eat from the Tree of Good and Evil, even his knowledge of evil would have become a seat, a foundation for his knowledge of good (*Toldot Ya'akov Yosef, Lech Lecha* 1).

▶▶ The *gematria* of "**foundation**" (יְסוֹד) is 80. The value of its first filling (יוד סמך וו דלת) is 586 and that of its second filling (יוד וו דלת) and (סמך מם כף וו דלת למד תו) is 1704. The sum of all three is 2370 = 5 · 474, where 474 is the *gematria* of "knowledge" (דַּעַת).

---

57. 48:13.
58. See *Zohar* II, 20a.
59. See *Likutei Torah, Kedoshim.*
60. Genesis 4:1.

root the world was created) and Sarah (בְּרִית אַבְרָהָם שָׂרָה). This is the only explicit marital union among the six covenants of fire.

The Arizal[61] explains that when husband and wife unite in marital relations the man enters the state of consciousness of the *sefirah* of victory, symbolized by the right leg, and his wife enters the state of consciousness of the *sefirah* of thanksgiving, symbolized by the left leg. There they unite and their flesh becomes one. In the context of Abraham and Sarah, during their marital union ▸ Abraham descends from his source in loving-kindness to assume the stance of victory (victory is the branch of loving-kindness, directly beneath it on the right axis of the Tree of Life) and Sarah ascends from her source in kingdom (the feminine principle) to assume the stance of thanksgiving.

To summarize:

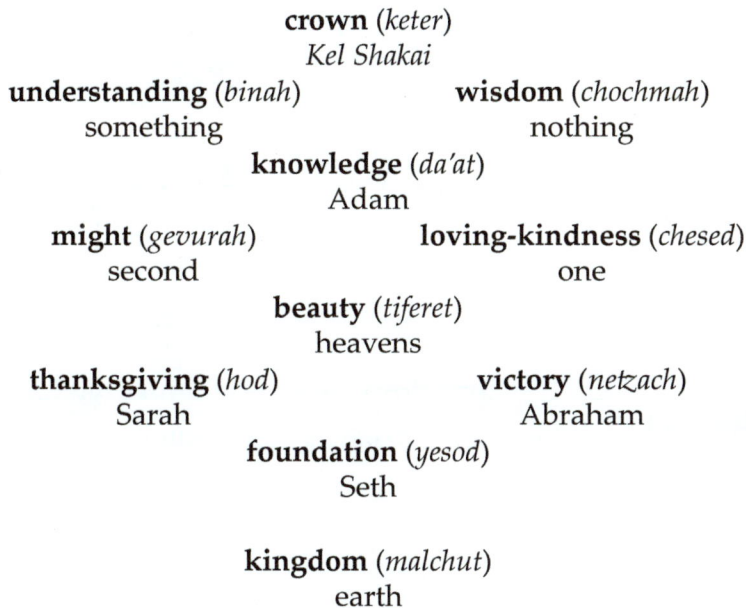

▸ **Marital union** brings new souls into the world. For most of their lives, Abraham and Sarah were barren. During the years that they were barren, Abraham and Sarah's marital union brought souls of converts into the world (*Megaleh Amukot parashat Vayeitzei*), the very converts that the Torah reports as, "the people they made in Charan" (Genesis 12:5).

crown (*keter*)
Kel Shakai

understanding (*binah*)　　　**wisdom** (*chochmah*)
something　　　　　　　nothing

knowledge (*da'at*)
Adam

might (*gevurah*)　　　**loving-kindness** (*chesed*)
second　　　　　　　　one

beauty (*tiferet*)
heavens

thanksgiving (*hod*)　　　**victory** (*netzach*)
Sarah　　　　　　　　Abraham

foundation (*yesod*)
Seth

kingdom (*malchut*)
earth

61. *Etz Chaim* 29:5.

Also by Rabbi Yitzchak Ginsburgh

The Hebrew Letters
Channels of Creative Consciousness
502 pages

Transforming Darkness into Light
Kabbalah and Psychology
192 pages

Living in Divine Space
Kabbalah and Meditation
288 pages

Anatomy of the Soul
144 pages

Awakening the Spark Within
Five Dynamics of Leadership that can Change the World
200 pages

Rectifying the State of Israel
230 pages

Kabbalah and Meditation for the Nations
216 pages

Lectures on Torah and Modern Physics
184 pages

The Mystery of Marriage
How to Find True Love and Happiness in Married Life
500 pages

Body, Mind and Soul
Kabbalah on Human Physiology, Disease and Healing
342 pages

The Art of Education
Internalizing Ever-New Horizons
302 pages

What You Need to Know About Kabbalah
190 pages

A Sense of the Supernatural
Interpretation of Dreams and Paranormal Experiences
208 pages

Consciousness and Choice
Finding Your Soulmate
284 pages

Frames of Mind
Motivation According to Kabbalah
256 pages

Gal Einai • www.inner.org

www.ingramcontent.com/pod-product-compliance
Lightning Source LLC
Chambersburg PA
CBHW060802150426
42813CB00059B/2863